Pollution

Other books in the Current Controversies series:

The Abortion Controversy
Alcoholism
Assisted Suicide
Biodiversity
Capital Punishment
Censorship
Child Abuse
Civil Liberties
Computers and Society
Conserving the Environment
Crime
Developing Nations
The Disabled
Drug Abuse
Drug Legalization
Drug Trafficking
Ethics
Europe
Family Violence
Free Speech
Garbage and Waste
Gay Rights
Genetic Engineering
Guns and Violence
Hate Crimes
Homosexuality
Hunger
Illegal Drugs
Illegal Immigration
The Information Age
Interventionism
Iraq
Marriage and Divorce
Medical Ethics
Mental Health
Minorities
Nationalism and Ethnic Conflict
Native American Rights
Police Brutality
Politicians and Ethics
Prisons
Racism
Reproductive Technologies
The Rights of Animals
Sexual Harassment
Sexually Transmitted Diseases
Smoking
Suicide
Teen Addiction
Teen Pregnancy and Parenting
Teens and Alcohol
The Terrorist Attack on America
Urban Terrorism
Violence Against Women
Violence in the Media
Women in the Military

Pollution

James Haley, *Book Editor*

Daniel Leone, *President*
Bonnie Szumski, *Publisher*
Scott Barbour, *Managing Editor*

CURRENT CONTROVERSIES

San Diego • Detroit • New York • San Francisco • Cleveland
New Haven, Conn. • Waterville, Maine • London • Munich

For more information, contact
Greenhaven Press
27500 Drake Rd.
Farmington Hills, MI 48331-3535
Or you can visit our Internet site at http://www.gale.com

LIBRARY OF CONGRESS CATALOGING-IN-PUBLICATION DATA

Pollution / James Haley, book editor.
p. cm. — (Current controversies)
Includes bibliographical references and index.
ISBN 0-7377-1187-6 (pbk. : alk. paper) — ISBN 0-7377-1188-4 (lib. : alk. paper)
1. Air—Pollution. 2. Water—Pollution. 3. Factory and trade waste—Environmental aspects. 4. Recycling (waste, etc.) 5. Environmental law. I. Haley, James, 1968– II. Series.
TD883 .P58 2003
363.73—dc21 2002024472

Printed in the United States of America

Contents

Foreword 10

Introduction 12

Chapter 1: Are Air and Water Pollution Serious Problems?

Chapter Preface 16

Yes: Air and Water Pollution Are Serious Problems

Air Pollution Threatens Human Health *by Pamela Meyer et al.* 17

Air pollution from both indoor and outdoor sources poses serious risks to human health. Efforts from government, business, and individuals are needed to further reduce the public's exposure to unhealthy air.

Factory Farming Is Polluting the Water Supply *by Andrew Nikiforuk* 26

Large-scale livestock farming, known as factory farming, is creating untreated animal waste that pollutes lakes, rivers, and other waterways. Factory-farm pollution threatens freshwater fishing and North America's supply of drinking water.

Coal-Fired Power Plants Create Harmful Emissions *by the Sierra Club* 31

Continued reliance on coal-fired power plants to generate electricity is producing smog, soot, acid rain, and other air pollutants that harm human health. These plants must be replaced with lesser-polluting means of producing electricity.

Chlorofluorocarbons Are Destroying the Ozone Layer *by Linda Baker* 36

The emission of chlorofluorocarbons (CFCs) from air conditioners and other sources is damaging the earth's ozone layer, which protects life from harmful ultraviolet radiation. The smuggling of CFCs into the United States must be stopped in order to halt the depletion of the ozone layer.

Automobiles Cause Air Pollution *by the American Council for an Energy-Efficient Economy* 44

Automobiles are polluting the environment by emitting carbon dioxide, airborne particles, nitrogen oxides, and other toxic chemicals. Overdependence on cars and trucks has contributed to the greenhouse effect and the threat of global warming.

No: Air and Water Pollution Are Not Serious Problems

Air Quality Has Improved *by the Environmental Protection Agency* 53
The quality of the air, as measured by concentrations of six major pollutants, has improved significantly during the past twenty years. In addition, emissions of five of the six major pollutants have decreased during this time period.

Factory Farming Is Not Polluting the Water Supply *by Dave Juday* 60
The Environmental Protection Agency has exaggerated the extent of water pollution from factory-farm livestock waste. Furthermore, scientific breakthroughs, such as breeding pigs that excrete less harmful waste, will make factory farming more environmentally safe in the future.

New Coal-Burning Technology Is Less Polluting *by Robert S. Kripowicz* 63
Advanced pollution control technology will enable the nation to use more coal as a source of electricity and still improve air quality. New coal-fired power plants are among the cleanest fossil-fuel power generating facilities in the world.

Air Pollution from Automobiles Has Been Reduced *by Joseph L. Bast and Jay Lehr* 70
Air concentrations of the six air pollutants tracked by the Environmental Protection Agency have fallen significantly in the last twenty years due to ongoing improvements in vehicle emission controls and tougher emission standards. As older cars and trucks are replaced with vehicles equipped with new pollution prevention devices, city air will become safer and cleaner well into the twenty-first century.

Chapter 2: Are Corporations Polluting the Environment?

Chapter Preface 77

Yes: Corporations Are Polluting the Environment

Corporations Are Reckless Polluters *by Russell Mokhiber* 78
Corporations have polluted the environment with chemical explosions, mercury poisoning, and oil spills. Public control must be reasserted over corporations to prevent further environmental degradation.

Corporations Are Engaging in Phony Environmentalism *by Richard Gilpin and Ali Dale* 81
Instead of changing business practices to reduce pollution, corporations have engaged in deceptive public relations campaigns to appear more environmentally friendly. The public has been "greenwashed" into believing that corporations care about the environment while companies continue to pollute.

Oil Companies Are Harmful Polluters *by Essential Action and Global Exchange* 85
Multinational oil companies in the oil-producing regions of Nigeria harm the environment and threaten public health. Companies employ the

cheapest practices for extracting oil, which have led to acid rain, pipeline leaks, and uncontained hazardous waste.

No: Corporations Are Reducing Pollution

Businesses Have Reduced Pollution *by Lynn Scarlett and Jane S. Shaw* 93

Businesses have made great strides in reducing pollution and conserving resources. Contrary to critics who blame corporate greed for endangering the environment, the profit-motive has encouraged businesses to engage in more efficient industrial activity, which results in less environmental destruction.

U.S. Factories in Mexico Are Reducing Pollution *by the Environmental Protection Agency* 98

American companies that operate factories in Mexico near the U.S. border—called "maquiladoras"—have implemented effective pollution prevention programs. As a result, several factories have made progress in reducing air and water pollution and hazardous waste, while conserving water, increasing recycling, and saving money spent on cleanup efforts.

Oil Companies Are Protecting the Environment *by the American Petroleum Institute* 101

The oil and natural gas industry has introduced new technologies at refineries and exploration and production operations that have substantially reduced the emission of pollutants. The industry has also taken steps to protect water environments by improving the designs of its ocean-going tankers and upgrading underground storage tanks.

Chapter 3: Are Pollution Regulations Effective?

Regulation and the Environment: An Overview *by Mary H. Cooper* 109

Politicians, business leaders, and environmentalists are at odds over whether government regulations can control pollution without slowing economic growth. Environmentalists are calling for new regulations to reduce air and water pollution, but opponents are not convinced such action is necessary.

Yes: Pollution Regulations Are Effective

The Superfund Hazardous Waste Program Is Effective *by the Environmental Protection Agency* 115

The federal regulatory program for cleaning up hazardous waste sites, known as "Superfund," has been successful in restoring contaminated land and reducing the health risks that toxic sites pose to communities. Without Superfund, thousands of unregulated hazardous waste sites would endanger human health and the environment.

Mandatory Pollution Regulations Are Effective *by Peter K. Krahn* 124

The results of a study conducted in British Columbia, Canada, demonstrate that industry often ignores voluntary pollution regulations. In contrast, when governments mandate and enforce environmental regulations, the industrial discharge of pollutants can be dramatically reduced.

Clean Water Regulations Should Be Strengthened
by Stephen B. Lovejoy and Jeffrey Hyde 131
Storm-water runoff carries pollutants, known as nonpoint-source pollution, from farms, mining operations, and urban areas into the nation's waterways. The Clean Water Act has reduced the discharge of industrial waste and sewage into surface waters, but more effective regulations are needed to prevent pollution from nonpoint sources.

Clean Air Regulations Protect Public Health *by Carol Browner* 137
Updated air quality standards that set tighter limits on ground-level ozone (smog) and fine particles (soot) will protect the public from the health hazards of breathing polluted air. Moreover, the new standards can be upheld without being too costly to industry.

No: Pollution Regulations Are Ineffective

The Superfund Hazardous Waste Program Is Ineffective
by Wayne T. Brough 141
The government's program for cleaning up hazardous waste sites, known as "Superfund," has become mired in legal proceedings that have diverted resources away from actual cleanups. As a result, cleanups have become more expensive and time-consuming and often do not reduce the human health risks posed by contaminated sites.

Clean Water Regulations Are Ineffective *by Richard A. Halpern* 151
The nation's regulatory system for preventing water pollution has not verifiably improved water quality. A more accurate assessment of the condition of the nation's waters is necessary before any clean water regulations can be effective.

Clean Air Regulations Are Ineffective *by Richard L. Stroup* 157
Regulations on the emission of hazardous pollutants are unnecessary and impose unfair costs on industrial firms. The Environmental Protection Agency routinely overestimates the health risks associated with air pollution, and it requires plants to install pollution control technology that has not been proven to benefit human health.

Chapter 4: Is Recycling an Effective Way to Reduce Pollution?

Chapter Preface 164

Yes: Recycling Is Effective

Recycling Is a Cost-Effective Means to Reduce Pollution
by Richard A. Denison and John F. Ruston 166
Antirecycling organizations have portrayed recycling programs as being too costly and of little benefit to the environment. Contrary to these assertions, recycling reduces dependency on increasingly expensive and polluting landfills for solid waste disposal. Recycling programs with high rates of participation have saved cities millions of dollars when compared to landfill costs.

Recycled Materials Produce Less Pollution *by Allen Hershkowitz* 174
Manufacturing products from virgin materials causes more air and water pollution than using recycled materials. Recycling also decreases pollution by reducing the need for mining, timber harvests, and petroleum refining.

Efficient Recycling Programs Can Reduce Waste *by Brenda Platt and Neil Seldman* 182
Recycling programs are an important component in reducing the pollution that results from the overconsumption of consumer products. Communities that embrace reuse, recycling, and waste reduction have experienced economic growth while at the same time making their economies more environmentally sustainable.

No: Recycling Is Ineffective

Recycling Programs Are Unprofitable and Unnecessary *by Christopher Douglass* 188
Curbside recycling programs grew in response to a false crisis in landfill capacity and the erroneous assumption that recycling reduces pollution. These programs are a losing venture for most cities because program costs cannot be recouped by selling recycled materials back to industry for remanufacture.

Mandatory Recycling Can Waste Resources *by Alexander Volokh and Lynn Scarlett* 195
Governments around the world have imposed arbitrary levels of recycling in an effort to increase the use of recycled materials. Because it is at times more profitable and efficient for businesses to use virgin materials in the manufacture of certain products, across-the-board recycling mandates can waste resources and are unlikely to reduce pollution.

Recycling Does Not Reduce Waste *by Peter Werbe* 202
Municipal recycling programs do not significantly reduce the amount of waste that is incinerated or deposited in landfills because most waste is generated by industrial operations, not households. In addition, recycling encourages the consumption of plastic products, even though only a small percentage of the plastic manufactured is actually recycled.

Organizations to Contact 206
Bibliography 210
Index 215

Foreword

By definition, controversies are "discussions of questions in which opposing opinions clash" (Webster's Twentieth Century Dictionary Unabridged). Few would deny that controversies are a pervasive part of the human condition and exist on virtually every level of human enterprise. Controversies transpire between individuals and among groups, within nations and between nations. Controversies supply the grist necessary for progress by providing challenges and challengers to the status quo. They also create atmospheres where strife and warfare can flourish. A world without controversies would be a peaceful world; but it also would be, by and large, static and prosaic.

The Series' Purpose

The purpose of the Current Controversies series is to explore many of the social, political, and economic controversies dominating the national and international scenes today. Titles selected for inclusion in the series are highly focused and specific. For example, from the larger category of criminal justice, Current Controversies deals with specific topics such as police brutality, gun control, white collar crime, and others. The debates in Current Controversies also are presented in a useful, timeless fashion. Articles and book excerpts included in each title are selected if they contribute valuable, long-range ideas to the overall debate. And wherever possible, current information is enhanced with historical documents and other relevant materials. Thus, while individual titles are current in focus, every effort is made to ensure that they will not become quickly outdated. Books in the Current Controversies series will remain important resources for librarians, teachers, and students for many years.

In addition to keeping the titles focused and specific, great care is taken in the editorial format of each book in the series. Book introductions and chapter prefaces are offered to provide background material for readers. Chapters are organized around several key questions that are answered with diverse opinions representing all points on the political spectrum. Materials in each chapter include opinions in which authors clearly disagree as well as alternative opinions in which authors may agree on a broader issue but disagree on the possible solutions. In this way, the content of each volume in Current Controversies mirrors the mosaic of opinions encountered in society. Readers will quickly realize that there are many viable answers to these complex issues. By questioning each author's conclusions, students and casual readers can begin to develop the critical thinking skills so important to evaluating opinionated material.

Current Controversies is also ideal for controlled research. Each anthology in the series is composed of primary sources taken from a wide gamut of informational categories including periodicals, newspapers, books, United States and foreign government documents, and the publications of private and public organizations. Readers will find factual support for reports, debates, and research papers covering all areas of important issues. In addition, an annotated table of contents, an index, a book and periodical bibliography, and a list of organizations to contact are included in each book to expedite further research.

Perhaps more than ever before in history, people are confronted with diverse and contradictory information. During the Persian Gulf War, for example, the public was not only treated to minute-to-minute coverage of the war, it was also inundated with critiques of the coverage and countless analyses of the factors motivating U.S. involvement. Being able to sort through the plethora of opinions accompanying today's major issues, and to draw one's own conclusions, can be a complicated and frustrating struggle. It is the editors' hope that Current Controversies will help readers with this struggle.

"Important progress has been made in cleaning up the country's air and water."

Introduction

The book *Silent Spring* by Rachel Carson, first published in 1962, awakened a passionate minority of environmentalists to the extent of the pollution problem in the United States. Carson chronicled the toll that decades of indiscriminate pesticide use, in particular DDT, which has since been banned, had taken on land, water, and human health. In an era when the environmental movement was still in its infancy, and notions of protecting the environment remained alien to much of the public, the pollution described in *Silent Spring* strongly affected many young readers. Former vice president Al Gore, who read the book as a teenager, recalls, "The publication of *Silent Spring* can properly be seen as the beginning of the modern environmental movement. For me personally, *Silent Spring* had a profound impact. . . . Rachel Carson was one of the reasons why I became so conscious of the environment and so involved with environmental issues." In the wake of *Silent Spring*, it seemed that the battle lines were officially drawn between the fledgling environmental movement and corporate leaders who, according to Gore, dubbed the book "hysterical and extremist."

As the 1960s wore on, a series of high-profile episodes of industrial pollution lent increasing authority to the environmental movement and its call for comprehensive pollution regulations. Nineteen sixty-nine proved to be a particularly rough year for the public relations departments of industrial polluters. In March of that year, an oil well operated by the Union Oil Company off the coast of Santa Barbara, California, blew out, covering more than four hundred square miles of ocean and thirty miles of beaches with bird- and fish-killing sludge. As Mary Graham comments in her book *The Morning After Earth Day*, "The spreading puddle of oil, from a well in the ocean floor that leaked for more than a week before company employees brought it under control, provided television viewers across the country with repeated reminders that government and industry had failed to prevent a disaster." Before the year was out, the public received yet another reminder that pollution was a serious problem, this time seeming to defy the laws of nature—the Cuyahoga River in Cleveland, Ohio, caught on fire. The river was so coated with oil and other flammables that it ignited when sparks from a passing train made contact with the river's surface. Graham quotes a Cleveland State University professor as saying, "The river 'burned on newscasts all over the world.'. . . It became 'a vivid symbol of the state of many of America's waterways.'"

Watching a river burn was a novel sight for American television viewers, and one that must not have inspired much confidence in politicians or corporate leaders. But in 1969, most Americans remained unconcerned about the environment. In fact, according to a White House poll conducted by Opinion Research of Princeton, New Jersey, in May 1969, one month prior to the Cuyahoga River fire, a mere 1 percent of the public expressed concern for the environment. Less than a year later, however, on April 22, 1970, the nation's first Earth Day demonstration galvanized tremendous public support for the environmental movement. The event was organized by Senator Gaylord Nelson of Wisconsin as a nationwide teach-in to inform Americans about the extent of pollution in their country, and what actions could be taken to prevent further environmental degradation. Earth Day 1970 turned out to be the largest organized demonstration in the nation's history, with over 20 million people participating in events held across the country. The enormous turnout meant that politicians could no longer ignore the environmental movement and its legions of new supporters. A second Opinion Research poll conducted in May 1971 showed that 25 percent of the public had grown concerned about the environment.

The response from government was startlingly swift and effective. President Richard Nixon submitted a plan to Congress on July 9, 1970, to reorganize the numerous agencies responsible for overseeing environmental programs into one department. The result was the formation of the Environmental Protection Agency (EPA) in December 1970 to develop and enforce the nation's pollution regulations. In the early 1970s, Congress gave the EPA plenty of new laws to enforce as it embarked on an unprecedented run of environmental legislation. Its first major achievement was the Clean Air Act of 1970, which directed the EPA to set national air quality standards to protect public health and the environment from industrial and automotive emissions. The Clean Water Act, which followed in 1972, established limitations on the amount of pollutants discharged by industry and city sewer and water treatment systems into rivers, lakes, and oceans. In 1976, the Resource Conservation and Recovery Act (RCRA) instituted a framework for the "cradle-to-grave" management of hazardous and nonhazardous wastes. The impact of these laws was soon felt by the numerous cities and businesses that were found in violation of the new clean air and water standards and consequently sued by the EPA.

Thirty years after Americans first awakened en masse to the pollution in their midst, it is clear that important progress has been made in cleaning up the country's air and water. Companies like General Electric (GE) can no longer dump carcinogenic chemical waste into rivers, as GE did for many years from its factories along the Hudson River. New technologies are enabling power plants to generate electricity while producing less harmful emissions. Urban areas have reduced smog and ground-level ozone through the phase-out of leaded gasoline and the introduction of cars and trucks that emit far less pollutants than those on the road just ten or fifteen years ago. The general perception is that the shocking pollution depicted by Rachel Carson in *Silent Spring* is largely a thing of the past, thanks to public concern and political action.

The reduction in obvious sources of pollution—the factory funneling toxins into the town river or the neighborhood hazardous waste site—may have lulled the public into a sense of complacency. The big culprits are largely under control, but pollution is also caused by less visible sources that are difficult—and costly—to contain. Examples include runoff from urban sewer systems, animal waste from farms, toxic gasoline additives leaking into groundwater, and ozone-depleting chlorofluorocarbons (CFCs) used for refrigeration. Because the sources of pollution have grown simultaneously more complex and less visible to the public, mustering support to further reduce pollution is difficult. Explains Jack Lewis, a former assistant editor with the *EPA Journal*,

> The challenges of the future involve extremely important but less visible problems of cross-media pollution, stratospheric ozone depletion, radon contamination, and protection of air and water supplies against ever-proliferating types of toxic chemicals in trace concentrations. . . . Unfortunately, in many cases, the public's evaluation of what most needs fixing . . . does not always square with expert scientific analyses of the most pressing dangers confronting [the public] . . . and their natural environment.

It may be that, much as in Rachel Carson's era, public awareness is lagging behind the increasingly complex problem of pollution. Rivers may no longer be catching on fire, but a dramatic image often serves to coax a distracted public into action. Although today's pollution problems lack the visible impact of the past, they should be taken no less seriously. *Pollution: Current Controversies* examines the debate over which sources of pollution are most threatening to human health and the environment and what measures business, government, and the public should take to reduce these threats.

Chapter 1

Are Air and Water Pollution Serious Problems?

Chapter Preface

In March and April 1993, thousands of Milwaukee, Wisconsin, residents became ill after the city's water supply was infected by an outbreak of *Cryptosporidium*, a parasite commonly found in the feces of dairy calves. Runoff from the spring thaw and heavy rains was thought to have carried contaminants from the Milwaukee area's many dairy farms into Lake Michigan, the city's water source. The outbreak caused the deaths of fifty people with weakened immune systems and sent thousands to area hospitals. The event drew attention to the problem of "nonpoint" sources of water pollution—that is, polluted runoff from farms, construction sites, city streets, sewer systems, and other sources—affecting up to half of the nation's waters.

In the thirty years since the passage of the Clean Air Act of 1970 and the Clean Water Act of 1972, growing populations and changing technologies have added new challenges to the prevention of air and water pollution. While pollution regulations—written with heavy industry in mind—have succeeded in forcing cities and industries to clean up their acts, regulations have proven less effective at controlling nonpoint water pollution and its growing and varied sources. Further complicating efforts to control pollution is the fact that activities traditionally regarded as nonindustrial, such as agriculture, have become as polluting as many industries yet remain unaffected by regulation.

For example, small family farms growing crops and raising livestock for meat and dairy products are being replaced by large-scale "factory farms" run by corporations. According to Mark Floegel, a journalist who writes about agricultural issues, "At the state-of-the-art Smithfield Packing Co. in Tar Heel, North Carolina, thirty-two thousand hogs per day are killed in the 973,000 square-foot plant." With so many animals in one place, waste containment and disposal are problematic. Explains Floegel, "Waste from industrial hog operations is captured in open pits . . . some of which can hold millions of gallons of semi-liquid excreta at any given moment. . . . But the waste pits are not always managed properly and environmentalists warn that accidents are bound to occur." Heavy rains, such as occurred during the Milwaukee outbreak of *Cryptosporidium*, can send that waste streaming into nearby rivers and lakes, threatening public health. However, proposed federal rules to regulate the discharge of farm waste in the same way that industrial waste is regulated have been criticized as too costly for small farmers, whose scale of operations does not approach that of factory farms.

Nonpoint water pollution generated by factory farms is one example of how complex the pollution problem has become. The viewpoints in the following chapter discuss which sources of air and water pollution pose the most serious threats and what prevention measures should be taken.

Air Pollution Threatens Human Health

by Pamela Meyer et al.

About the author: *Pamela Meyer is an epidemiologist working for the air pollution and respiratory health branch of the National Center for Environmental Health in Atlanta, Georgia.*

In the course of a day, we breathe 5,000 to 15,000 liters of air. With each breath, we inhale life-sustaining oxygen, which is absorbed in our lungs and carried throughout our body. Air also contains pollutants, including pollen, microbes, particles such as soot and dust, and gases such as carbon monoxide—substances that can harm the human body. Contact with these harmful substances, which are filtered through the lungs and can also irritate the eyes and skin, triggers several defense mechanisms such as coughing, sneezing, and the production of secretions. When these defense mechanisms are overwhelmed, human tissue is damaged or destroyed. Chronic or severe exposure may hasten the onset and progression of disease and even result in death.

The Need for Air Pollution Standards

Although air pollution has plagued crowded cities for centuries, several episodes in the United States and Europe since 1930 have driven home the harmful effects of air pollution. The worst air pollution episode in the United States occurred in Donora, Pennsylvania, on October 26, 1948, when sulfur dioxide, carbon monoxide, and metal dust emitted by local zinc smelter smokestacks was trapped by stagnant air and formed poisonous compounds over the industrial town. During the next five days, 43 percent of the 14,000 people in the community became sick. Ten percent of them were severely affected, and 19 people died. Statistically, only two deaths would have likely occurred in that small a population at the time.

Perhaps the most severe episode of ambient air pollution in the world occurred in London, England, in December 1952, when stagnant air trapped thick fog and air pollution for several days. More than 4,000 excess deaths were recorded. These acute episodes motivated the United States and other countries to implement ambient air-quality standards and strategies to reduce emissions that contribute to air pollution.

Since the implementation of these standards, air pollution levels have decreased in many parts of the world, but current ambient concentrations still cause adverse health effects. In particular, air pollution exacerbates chronic heart and lung disease and causes death. Although the most common cause of heart and lung illness and death in the United States is tobacco smoke, there is substantial evidence of the harmful effects of air pollution. One way to reduce our risk of illness from air pollution is to learn about the common air pollutants so we can control our exposure to them.

Congressional Action and Outdoor Air Pollution

Outdoor air pollution is produced primarily by the burning of fossil fuels by motor vehicles, power plants, and industries. Concern about reduced visibility and evidence of adverse health effects led Congress to enact several laws concerning air quality. Beginning in 1955, air pollution research was authorized by the Air Pollution Control Act. Later, the 1963 Clean Air Act authorized the federal government to legislate and enforce air pollution controls. Paving the way for national air quality standards was the Motor Vehicle Air Pollution Act of 1965, which defined a process for implementing national emissions standards for new motor vehicles. But the 1970 Clean Air Act established the public health basis of the nation's effort to control air pollution.

Subsequently, Congress established the U.S. Environmental Protection Agency (EPA) and charged it with setting National Ambient Air Quality Standards (NAAQS) to protect the public's health, including the health of sensitive groups within the population. EPA's role is to identify air pollutants that are likely to endanger public health. Accordingly, EPA identified six air pollutants—known as the criteria pollutants—which pose the greatest threat to our health: ozone, sulfur dioxide, particulate matter, nitrogen dioxide, carbon monoxide, and lead. The Clean Air Act of 1990 also charges EPA with periodically reviewing and, if appropriate, revising the NAAQS to keep standards in line with current scientific knowledge.

"Air pollution exacerbates chronic heart and lung disease and causes death."

After the United States phased out tetraethyl lead—a highly toxic additive that took the knock out of automotive engines—from gasoline in the mid-1970s to 1980, airborne lead levels decreased, and more importantly, blood lead levels among children in the nation also decreased. From 1988 through 1997, ambient lead concentrations decreased 67 percent. While lead from paint in older homes continues to pose a health threat, especially to young children, lead is no longer considered a major source of outdoor air pollution in this country.

Ozone. Ozone occurs naturally in the stratosphere, seven to 31 miles above the Earth, and protects human health by blocking the sun's harmful ultraviolet rays. In contrast, ground-level ozone is produced by chemical reactions with nitrogen dioxide and volatile organic compounds such as benzene and toluene in the atmosphere, and it is the main component of smog. Because the formation of ground level ozone is stimulated by sunlight and heat, ozone levels peak in late spring and summer and during the afternoon—when people spend more time outdoors.

Ozone is a powerful respiratory irritant that can interfere with the lung's immunity, constrict airways, and increase respiratory symptoms in healthy adults and susceptible people. Most vulnerable are the very young, whose lungs are immature; the elderly, whose lungs are less effective at filtering irritants; those with lung disease such as asthma and emphysema, and those with heart disease. While the adverse effects of short-term exposure to ozone are well documented, researchers are conducting studies of the long-term effects of repeated, intermittent exposures to ozone.

"Ozone is a powerful respiratory irritant that can . . . constrict airways and increase respiratory symptoms in healthy adults and susceptible people."

Particulate matter. Particulate matter includes naturally occurring dust and pollen as well as soot and aerosols from combustion activities such as agricultural burning, transportation, manufacturing, and power generation. The most harmful particles are not the large particles, which are mostly removed in the upper airways, but the small particles that may be deposited deep in the lungs. Before 1987, the standard for measuring particulate matter was based on total suspended particulate matter, no matter the size. In 1987, EPA changed its standard to measure only the percentage of particles with an aerodynamic diameter of 10 microns or less. However, recent research has shown that fine particulate matter—which includes particles with an aerodynamic diameter smaller than 2.5 micrometers—is inhaled deeper into lung tissue, and is therefore more harmful. In 1997, EPA issued new standards to address these smaller particles, which several epidemiological studies have linked with decreased lung function, increased respiratory symptoms, increased school absenteeism, increased respiratory hospital admissions, and increased mortality, especially from respiratory and cardiovascular failure.

In contrast to controlled laboratory studies, epidemiological studies measure human health effects of exposure to ambient air pollution. Ambient air typically contains several pollutants, and epidemiological studies allow researchers to evaluate the effects of individual and combined pollutants. Since epidemiological studies are observational, it is possible to study the health effects among vulnerable populations.

Sulfur dioxide. The burning of sulfur, a natural contaminant of all fossil fuels, results in the formation of sulfur oxides. Sulfur dioxide is produced primarily by industrial and electrical power-generating processes involving fossil fuel combustion. Sulfur dioxide combines with atmospheric water, oxygen, and oxidants to create weak acids that fall to the Earth as dry particles, snow, fog, or rain, which is commonly referred to as acid rain. When these acidic substances fall to the Earth, they can harm vegetation and acidify lakes and streams. Sulfur dioxide can also constrict air passages, making breathing difficult for those with asthma, and may also alter the immune system and aggravate existing cardiovascular disease.

"Air pollution, once viewed as a local problem, especially in urban areas, has become a regional issue."

Nitrogen dioxide. Nitrogen dioxide is a product of high-temperature combustion and contributes to the formation of ozone. Motor vehicle emissions are the primary source of nitrogen dioxide in outdoor air, but power plants and fossil-fuel-burning industries also contribute. Nitrogen dioxide can irritate the lung and alter its defense mechanisms, thereby increasing a person's risk for respiratory infections.

Carbon monoxide. Carbon monoxide is produced during the incomplete combustion of carbon-containing materials, including gasoline, natural gas, oil, coal, wood, and tobacco. The principal source of carbon monoxide in outdoor air is motor vehicle emissions. Outdoor concentrations of carbon monoxide vary depending on how and where and when the gas is produced. For example, in urban areas, carbon monoxide levels are greatest in downtown areas where motor vehicle density is high, during peak commuting times, and in the passenger compartments of motor vehicles. Carbon monoxide interferes with the ability of the blood to carry oxygen to tissues; the most sensitive of these tissues are in the heart and brain. The health effects of carbon monoxide poisoning range from impaired mental alertness and performance, headaches, nausea, fatigue, and dizziness to coma and death.

Outdoor Control

Strategies to reduce outdoor air pollution include implementing automobile emission standards, improving technology to reduce smokestack emissions of particulate matter, and requiring more-stringent standards for sulfur content in fossil fuels. Levels of the six criteria pollutants all decreased from 1988 to 1995. The greatest decrease was for lead, at 67 percent, and the least was for nitrogen dioxide, at 14 percent.

Air pollution, once viewed as a local problem, especially in urban areas, has become a regional issue. Sulfur dioxide, particulate matter, and the precursors of ground-level ozone can travel long distances. Industries contributed to the problem when they switched from short smokestacks to tall smokestacks, which released pollutants at higher levels in the atmosphere where they could be transported longer distances and cross geopolitical boundaries.

Several regional organizations have been created to address regional air pollution issues in the United States. These organizations vary in the composition of their members, but many include representatives from federal, state, and local agencies; environmental groups; industry; academic institutions; and private citizens.

The 1990 Clean Air Act Amendments, for example, established the Ozone Transport Commission and the Northeast Ozone Transport Region to address long-standing ozone problems in the northeastern United States. Commission representatives include governors and air pollution-control officials from each of the 12 members states—Maine, New Hampshire, Vermont, Massachusetts, Rhode Island, Connecticut, New York, Pennsylvania, New Jersey, Delaware, Maryland, and Virginia—and the District of Columbia. Administrators from three northeastern EPA Regions also participate. To reduce regional air pollution, the members have agreed to introduce a low-emission vehicle program and to reduce emissions of nitrogen oxides.

Aside from mandated organizations, there are also voluntary organizations whose mission is to find regional solutions to regional problems. For example, the Southern Appalachian Mountains Initiative, which is led by eight southern states in the Appalachian region, works with EPA, industries, federal agencies, academic institutions, environmental groups, and private citizens to seek solutions to the region's specific challenges. Because of the geography and meteorological conditions of the area, air pollution tends to stagnate over the area, which includes 10 of the nation's national parks and wilderness areas. . . .

"Only recently have indoor sources of air pollution received much attention."

A larger regional group that works to address long-range transport of air pollution is the Ozone Transport Assessment Group (OTAG), which was formed to identify and recommend cost-effective control strategies for volatile organic compounds and nitrogen oxides to facilitate compliance with NAAQS for ozone. OTAG is a partnership between EPA and the Environmental Council of States and includes representatives from 37 states east of the Rocky Mountains, industry, and environmental groups.

In addition to regional groups within the United States, there are international agreements with Mexico and Canada to control air pollution. Created in 1994 under the North American Free Trade Agreement (NAFTA), the Commission for Environmental Cooperation addresses air pollution control in the three countries to ensure that pollution created in one country does not affect the health of citizens in another.

The creation of these regional and international cooperative groups is evidence of increased attention being paid to the necessity of addressing air quality issues across arbitrary boundaries. In addition, there is a growing consensus that a strict regulatory approach alone is inadequate to address these problems.

Indoor Air Pollution

While we've spent decades working to clean up the air we breathe outside, only recently have indoor sources of air pollution received much attention. Since the oil crisis of the 1970s, office and home construction of new buildings and retrofitting of old buildings have created airtight structures. In addition, new materials such as particle board and carpet can contain high levels of chemicals that are trapped inside and are emitted into the air long after installation.

In recent years, EPA and its Science Advisory Board ranked indoor air pollution among the top five environmental risks to public health. Indoor air pollutants probably have a greater impact on our health than outdoor pollutants because people in the United States tend to spend more time indoors than outdoors.

Tobacco smoke and emissions from unvented combustion appliances, woodstoves, and fireplaces are the principal indoor air pollutants; other potential pollutants include biologic agents such as bacteria and viruses, naturally occurring carcinogenic radon, dusts, and volatile organic compounds found in office and home furnishings.

Outdoor air pollutants may also enter a building through open windows or ventilation systems and contribute to the concentration of indoor air pollutants; the degree of infiltration depends on the characteristics of a building's construction and the efficiency of its heating, ventilation, and air conditioning system.

Moreover, while workers in factory and construction jobs are protected through the Occupational Safety and Health Administration from occupational hazards such as exposure to toxic emissions, no single federal agency has statutory jurisdiction over indoor air quality. The responsibility for indoor air quality research and policy . . . is shared by several federal agencies. The EPA established a research program to address radon and other indoor pollutants; the Department of Housing and Urban Development sets standards for agency-funded projects and for mobile homes; the Consumer Product Safety Commission regulates injurious products that pollute indoor air, such as asbestos; and the Department of Energy has supported the development of more-efficient and less-polluting energy technologies, and it monitors the health effects of energy conservation.

"Secondhand tobacco smoke . . . has been associated with low birth weight, sudden infant death syndrome, and acute lower respiratory tract infections among children."

Federal efforts to reduce indoor air pollution include developing voluntary industry codes, establishing product safety standards, publishing guidelines for dealing with radon, and offering guidance for handling asbestos in schools.

Therefore, it is important for building supervisors in schools and office buildings, and individual homeowners, to educate themselves on the possible sources of indoor air pollution and to work toward reducing exposure to occupants.

Of the many possible sources of indoor air pollution, six pollutants are of particular concern in terms of public health.

Primary Indoor Air Pollutants

Combustion by-products. Incomplete combustion of wood and fossil fuels such as coal, oil, and gas produces nitrogen oxides, carbon oxides, and particulate matter. The concentrations of combustion products in our homes depend on the efficiency of combustion and ventilation and on the maintenance and function of heat-generating equipment. Gas stoves, which produce nitrogen dioxide and carbon monoxide, are used by half of the U.S. population. The use of gas stoves for cooking in homes has been linked to an increased risk for lower respiratory illness among children. Gas or kerosene space heaters emit carbon monoxide, nitrogen dioxide, and particles, and these fuels contain high levels of sulfur. In addition, each year hundreds of people die from carbon monoxide poisoning in homes, automobiles, and other enclosed spaces with improper ventilation.

"Asbestos . . . [once] commonly used in building construction . . . causes lung diseases."

Tobacco smoke. Tobacco smoke contains more than 4,500 compounds, 50 of which are known or suspected carcinogens, and six of which are developmental or reproductive toxicants. The undeniable health effects of primary cigarette smoking include premature mortality, lung cancer, and obstructive lung diseases such as emphysema. Secondhand tobacco smoke, or environmental tobacco smoke, has been associated with low birth weight, sudden infant death syndrome, and acute lower respiratory tract infections among children. Secondhand smoke can also aggravate asthma, and it is associated with acute and chronic heart disease as well as mortality from heart disease.

Volatile organic compounds. Volatile organic compounds—gases that occur at normal temperatures from a wide variety of human made products—are emitted by modern furnishings, construction materials, and consumer products. One of the most common of these compounds is formaldehyde, which is used in many products commonly found in homes, such as cosmetics, toiletries, and the resins used in laminated wood products and particle board. Harmful vapors can be emitted for long periods after these materials are installed. For example, urea formaldehyde foam insulation, which became popular in the mid-1970s, emits a burst of formaldehyde immediately after application and then continuously emits lower levels. When improperly installed, formaldehyde can be released at high concentrations indoors. Formaldehyde irritates the respiratory tract and at high concentrations is toxic.

Asbestos. From the beginning of the century until the early 1970s when EPA banned its use in certain applications, asbestos was commonly used in building construction for thermal and acoustic insulation and fire protection. Asbestos causes lung diseases, especially a chronic irritation and inflammation of the lung, asbestosis, but also lung cancer and mesothelioma—a malignant tumor of the lining of the lung—among people exposed to asbestos in the workplace. Whether people in nonoccupational settings are at risk for lung cancer has not yet been resolved with certainty. Although asbestos use has declined in the United States, asbestos-containing materials are still present in many homes, schools, and offices.

Radon. Radon is a radioactive gas created during the decay of radium, which itself is a decay product of naturally occurring uranium. Natural radon gas in the soil is the main source of radon in buildings and can penetrate through the foundation into the air in homes. EPA estimates that as many as 6 million homes throughout the country have elevated levels of radon. Elevated radon concentrations can cause lung cancer.

Biologic contaminants. Biologic contaminants, which are present to some extent in all buildings, can become airborne and enter our respiratory systems, causing infections and disease. They can also trigger allergic reactions and asthma attacks. Such contaminants include pollens; house dust mites; insect excreta and body parts; animal dander and excreta; and microbes such as viruses, bacteria, fungal spores, protozoans, and algae. Biologic contaminants can be found in any environment that provides nutrients and moisture for their growth.

Indoor Control

The sources of indoor air pollutants are diverse and require different control measures. Control of environmental tobacco smoke, one of the most common and harmful indoor pollutants, can be accomplished by limiting areas where people can smoke. Employer and government policies have been successful in decreasing secondhand smoke in work sites and public areas, but these policies obviously cannot be enforced in private homes.

The presence of asbestos in a home or building does not necessarily indicate risk to health. Asbestos becomes harmful when it is damaged or disturbed and its fibers become airborne. Encapsulating asbestos by applying sealants to surfaces or removing it may reduce the risk of exposure.

Because radon can cause lung cancer, it is important to test for the presence of radon. Homeowners can purchase low-cost radon test kits or hire a trained contractor to test for radon. If high levels of radon are found, remediation may be necessary. This generally requires sealing a building's foundation to prevent soil gases from entering, or venting the gas produced underneath the foundation to the outside of the building.

Strategies for the control of indoor biologic contaminants include reducing relative humidity; repairing leaks and seepage from roofs and water pipes;

properly maintaining heating, ventilating, and air conditioning equipment; and cleaning buildings regularly and avoiding the use of toxic cleaners.

Despite improvements in air quality, nearly one in five Americans, or 50 million people, lived in counties that exceeded the NAAQS for at least one pollutant in 1996. Because of the considerable number of people still exposed to air pollution, we need continued evaluation of the safety of current standards. We also need to incorporate new information into regulations to control air pollution, as EPA did in 1997 by recommending more stringent standards for ozone and fine particulate matter, effectively doubling to 107 million the number of people living in polluted areas.

"Because of the considerable number of people still exposed to air pollution, we need continued evaluation of the safety of current standards."

We also need support from the health community. A goal for reducing the public's exposure to harmful air has been established as part of Healthy People 2000, a national prevention initiative. For two decades, the U.S. Department of Health and Human Services has used health promotion and disease prevention objectives to improve the health of the American people.

The Healthy People 2000 objective for [decreasing the public's exposure to] air pollution is to increase the proportion of people who live in counties that have not exceeded any EPA standard for air quality in the previous 12 months to 85 percent by the year 2000. [The Healthy People 2000 initiative met 21 percent of its targets in 2000.] In 1996, 81.3 percent of the population lived in counties that did not exceed any EPA standard in the previous 12 months, a substantial increase from 49.7 percent in 1988. Tighter standards, though, could decrease this percentage to the 60 to 65 percent range.

Strides in improving air quality must continue, particularly to protect people most susceptible to the adverse effects of both indoor and outdoor air pollutants, such as children; the elderly; tobacco smokers; and people with pre-existing cardiopulmonary diseases, including asthma, allergic rhinitis, cystic fibrosis, and acquired immunodeficiency syndrome (AIDS).

Children, the largest susceptible group, spend more time than adults engaged in vigorous activities, and therefore have a higher relative intake of pollutants into their lungs. Children also spend more time outdoors than adults, particularly in the summer when ozone levels are highest.

Air pollution, whether indoors or outdoors, adversely affects human health. The effective control of air pollution will involve multiple approaches. Government can develop and enforce regulations to reduce ambient pollutants and environmental tobacco smoke, employers can encourage employees to carpool or use public transportation, and individuals can learn about air pollutants and make personal lifestyle changes to reduce their exposures. After all, improving air quality is everyone's responsibility.

Factory Farming Is Polluting the Water Supply

by Andrew Nikiforuk

About the author: *Andrew Nikiforuk is a journalist based in Calgary, Canada, who has written extensively about factory farms.*

Long after the dead have been buried in Walkerton, Ontario, rural Canadians who rely on groundwater will continue to feel and smell the impact of a largely unreported revolution: the growth of factory farms. This new industry, or what governments call "intensive livestock operations," has unsettled farm communities from New Brunswick to Alberta. Unlike the family enterprises of old, which proudly cared for 20 pigs or 60 cattle, these new facilities operate on an entirely different and largely unregulated scale.

The Rise of "Feedlot Alley"

Let's begin with the industrialization of Alberta's fabled beef herds. Thirty years ago, thousands of farmers throughout the province regarded the care of 100 cattle as a big deal. Today, 50 beef barons, largely concentrated north of Lethbridge in an area known as Feedlot Alley, fatten and manage 80 per cent of the province's slaughtered beef. As a result, just one feedlot will have as many as 25,000 cattle in a maze of outdoor corrals on a piece of land the size of a city block. As Cor Van Ray, Alberta's undisputed feedlot king, puts it, "Everyone likes to think they can get their chicken or beef on a cozy farm somewhere. But unless you get big and run it like a business you are squeezed out. This whole corporate thing is just snowballing."

Factory farming has also radicalized the country's multibillion-dollar hog industry in Ontario, Quebec and the West. One of the fastest growing in the world, Canada's hog sector employs 100,000 people and exports more than a third of its production to 35 countries. In 1976, 18,622 Ontario farmers raised an average of 103 pigs each. By 1996, 6,777 (many of them white-suited swine technicians) managed an average of 418 animals each in crowded high-tech barns, while just two per cent of Ontario's hog factories accounted for nearly a quarter of the 5.6 million hogs produced in the province. And big just keeps getting bigger. An Asian firm, the Taiwan Sugar Company, for example, proposes to build an 80,000-hog operation outside Lethbridge. Local citizens are concerned about the amount of untreated waste it will create—equivalent to that produced by 240,000 people. They are also concerned that, like most of Alberta's intensive livestock operations, it will be regulated and taxed like a family farm.

Threats to Water Quality

The monstrous size of these profitable operations has raised troubling questions about water quality and threats to public health from coast to coast. Manure from factory farms often contains a variety of heavy metals, lake-choking nutrients and deadly pathogens such as E. coli 0157. In fact, wherever factory farms have concentrated industrial piles of manure in small spaces, big trouble has followed. No one knows this better than Dr. Paul Hasselback, the medical officer of health for Alberta's Chinook Health Region, home to Feedlot Alley and the nation's largest concentration of livestock—and a region plagued by chronic health and water problems. "Walkerton has demonstrated to the public that there is a substantial risk out there," he notes. "There just isn't a framework to develop these industries in a sustainable fashion."

The market forces now erecting animal factories across Canada are simple. They include a federal commitment to support low food prices and new economic realities. For starters, it is far cheaper to export steak and pork than to ship grain or corn. Thanks to abundant feed grains, Western Canada can now produce bacon more profitably than any other region in the world. In addition, the world's key pork producers, Taiwan and Holland, recently pushed production into the danger zone, causing severe water pollution and animal disease outbreaks. But their environmental disasters have had an effect here: hog barns managed by Europeans or funded by Asian investors are popping up all over the country.

Such factories, however, have generated intense opposition in rural Canada. Living next to one can be unpleasant: in addition to the stench of manure, neighbours routinely complain about increased traffic, flies, dust and noise. Most Canadian provinces now boast some kind of coalition battling beef feedlots or hog barns—and the resistance generally focuses on fears about water pollution. And for good reason. The growth of animal factories—aided by provincial incentives such as subsidies in Quebec and the Prairie provinces—has created industrial-scale waste problems. A single 500-sow farm producing 20 piglets per sow a year creates as much effluent as a town of 25,000 people without a waste treatment system.

> ***"Alberta's livestock industry may hold a national manure record: dung heaps equivalent to the waste of 48 million Canadians."***

Hog waste, which contains a host of heavy metals because of mineral-rich feeds, simply goes to open-air lagoons before it is sprayed on the land. Beef factories aren't much better. A 25,000-head feedlot produces in excess of 50,000 tonnes of dung—or more fecal matter than 250,000 Calgarians excrete over a year. It, too, is just spread on land bases often too small to absorb all the nutrients. Alberta's livestock industry may hold a national manure record: dung heaps equivalent to the waste of 48 million Canadians. Very little of this dung is properly treated, regulated or monitored. In Alberta, to the dismay of public health officers like Hasselback, in May 2000 the provincial government unceremoniously shelved proposed legislation to crack down on and monitor intensive livestock operations. In many provinces, government downsizing has also foisted the responsibility for regulating these facilities on those least equipped to do the job: municipal governments.

In Quebec, where, according to government statistics, probably a third of all hog operations don't comply with provincial environmental standards, a coalition of 18 farm and environmental organizations even took their case to NAFTA's Commission for Environmental Co-operation. The governments of Mexico and Canada, however, voted against investigating allegations that Ottawa and Quebec weren't protecting waterways from manure runoff. Ontario is also in bad shape. Dr. Murray McQuigge, the outspoken public health officer who blew the whistle on the Walkerton outbreak, warned in September 1999 that "poor nutrient management on farms is leading to the degradation of the quality of groundwater, streams and lakes." Ontario has no specific legislation governing factory farms.

"Pollution from agriculture and other sources . . . will end all freshwater fishing within 50 years."

Les Klapatiuk, who runs a Calgary firm specializing in water treatment, says there isn't a single government in Canada with adequate legislation to deal with these volumes of animal waste. "The leakage from lagoons is incredible, and when you spread millions of gallons of waste on a field it just runs into the surface water," he says. "If a city or an oil company operated this way, they would be shut down."

Compromising Public Health

All this manure has already taken a costly toll on waterways in Quebec, Ontario, Manitoba and Alberta. A 1998 federal study found half of 27 Alberta streams in key agricultural production areas exceeded water guidelines for nitrogen, phosphorus and disease-carrying bacteria. According to a 1991 study, about 30 per cent of rural wells in Ontario were contaminated with pathogens. In the United States, the Environmental Protection Agency estimates that agricultural runoff from animal factories and traditional farms is the leading source of water pollution in that country.

David Schindler, one of the world's leading experts on water and an ecologist at the University of Alberta in Edmonton, believes Canada is no different. He thinks the nation's notoriously cavalier attitude towards water quality will prove calamitous. In a scientific paper to be published this fall, he predicts that pollution from agriculture and other sources, as well as habitat destruction, will end all freshwater fishing within 50 years, while the nation's drinking water supply will be in dire straits within a century. "Whenever you don't pay attention to factory farms and their waste, you end up paying for it in spades in health services and waste-water treatment," Schindler says. "Country after country has gone down this path. Why aren't we learning from other people's mistakes?"

"[Twenty-five] per cent of all workers employed by hog barns suffer from bronchitis due to the corrosive nature of hog waste."

Is health being compromised? In a study published in 1999, Health Canada mapped cattle densities and the incidence of Escherichia coli 0157 infections in rural Ontario, only to discover that six rural Ontario counties with the highest number of cattle—and Walkerton is located smack dab in the middle of them—routinely registered the highest rates of E. coli 0157 infection between 1990 and 1995. Pascal Michel, the Health Canada veterinarian and epidemiologist who did the E. coli 0157 study, says he was surprised by the scale of the Walkerton tragedy—but not by its location. "We knew we could expect more cases of infection in these counties than anywhere else in the province," he said.

Alberta's Feedlot Alley, which produces untreated waste from 1.3 million animals that is the sewage equivalent for a population of eight million people, has also been plagued by Walkerton-like troubles. Conclusive proof that health problems there are the result of animal waste does not exist. But area residents routinely run to the bathroom with the highest rates of intestinal disease in the province. In one three-year period between 1989 and 1991, E. coli 0157 killed a dozen children and afflicted scores more in southern Alberta's cattle country. In recent years, the Chinook Health Region has repeatedly raised pointed questions about the bacterial contamination of drinking water, the fouling of irrigation canals, clogged water treatment plants and nitrates in the groundwater.

The public health costs of hog factories are equally daunting. A U.S. survey published in spring 2000 found that people living downwind from hog farms in North Carolina—where such factories first originated—experienced more headaches, runny noses, sore throats, excessive coughing, diarrhea and burning eyes than residents of a community without hog factories. None of this is surprising: according to other U.S. studies, 25 per cent of all workers employed by hog barns suffer from bronchitis due to the corrosive nature of hog waste. A 1997 Iowa study found that the methane, ammonia and hydrogen sulfide spewing from a 4,000-hog operation caused respiratory illnesses in people living up to two kilometres away.

Instituting Reforms

In the United States, where factory farms have polluted parts of the eastern seaboard and poisoned scores of communities, state and federal governments have gotten tough. Kansas and Nebraska, for example, have banned large animal factories and Iowa has declared a moratorium on future developments. The Environmental Protection Agency (EPA) has also targeted factory farms for top priority inspections. Canada, however, hasn't followed suit. With the exception of a pending national program for uniform standards for hog operations, and funding on manure research, Ottawa is largely absent from the debate over factory farms. Nor are provinces picking up the slack.

Critics agree there are some obvious reforms. Provincial governments should cap livestock density in many regions, while many rural Canadians want to see animal factories regulated and taxed for what they are: industries. Canada also needs laws that recognize that E. coli 0157 and other pathogens have forever changed the nature of manure. Many experts also recommend that animal waste should be properly treated before the dung ever leaves the barn. Most producers support higher standards for the simple reason that disasters like Walkerton aren't good for business. Last but not least, Schindler, Canada's top water scientist, would also like to see federal funding for freshwater research restored (it is now, he says, at an all-time low) and comprehensive management plans for the nation's watersheds. "Walkerton," Schindler concludes, "should be a wake-up call—for the entire nation."

Coal-Fired Power Plants Create Harmful Emissions

by the Sierra Club

About the author: *The Sierra Club is an environmental group with over seven hundred thousand members that works to protect the natural and human environment.*

We all use electricity in our daily lives, almost without thinking about it—turning on the lights, listening to the radio, and using computers. If we stopped and learned about the energy we use, we would encounter some shocking realities about the impacts of the energy production process on the environment and our health.

Reliance on Coal Generates Pollution

With all the amazing technological advancements over the last century, one thing that has not changed very much is our reliance on fossil fuels, in particular, dirty coal to generate electricity. In the U.S. today, coal is the number one source of electricity produced (54%), followed by nuclear sources (21%), hydropower (16%), natural gas (9%), oil (2%), and other non-renewables (3%). As the producer of the largest share of our nation's energy, coal-fired plants are also some of the dirtiest.

Many older coal-fired power plants have enjoyed a loophole in the Clean Air Act, allowing them to avoid modernizing with pollution controls. As a result, as many as 600 existing power plants are between 30–50 years old and are up to 10 times dirtier than new power plants built today. When the Clean Air Act was proposed, this loophole was included to get it passed because Congress assumed that newer plants would come into compliance with the Clean Air Act standards and soon replace the older more polluting plants. For a variety of reasons, including efforts to heavily subsidize coal, this has not happened. Therefore, we are now faced with a disproportionate amount of pollution coming from these old, dirty, under-controlled plants.

From "Clean Air: Dirty Coal Power," by the Sierra Club, www.sierraclub.org, September 12, 2001.

Out of the entire electric industry, coal-fired power plants contribute 96% of sulfur dioxide emissions, 93% of nitrogen oxide emissions, 88% of carbon dioxide emissions, and 99% of mercury emissions.

Smog

When nitrogen oxide (NO_x) reacts with volatile organic compounds (VOCs) and sunlight, ground level ozone, or smog forms. Power plants are second only to automobiles as the greatest source of NO_x emissions. NO_x emissions from huge dirty coal plants with tall smokestacks in the midwest are often blamed for increased smog levels in many eastern regions because smog and its precursor pollutants are easily transported hundreds of miles downwind from pollution sources. More than 100 million Americans live in regions that fail to meet health-based smog standards.

Even our national parks have not escaped the smog caused by coal-fired power plants. Regional haze from airborne pollutants has reduced annual average visibility in the U.S., to about one-third in the west and to one-quarter in the east, of natural conditions. Smog concentrations increased at 17 of 24 National Park Service monitoring sites from 1992–1998. In fact, . . . Cape Cod National Sea Shore has had higher pollution levels and more bad air days than Boston, and Shenandoah National Park in Virginia recorded higher levels of smog than any city in the Southeast except Atlanta, GA, and Charlotte, NC, in 1998.

> ***"As the producer of the largest share of our nation's energy, coal-fired plants are also some of the dirtiest."***

When inhaled, smog causes a burning of the cell wall of the lungs and air passages. This eventually weakens the elasticity of the lungs, making them more susceptible to infections and injury and causing asthma attacks and other respiratory illnesses. This danger is present for anyone who inhales smog, although children, the elderly, and those with respiratory problems are at a higher risk of developing health problems associated with smog pollution. A UCLA School of Medicine study found that over time, repeated exposure to smog and other air pollutants can cause as much damage to the lungs as smoking a pack of cigarettes a day. In addition, a recent Abt Associates study found that high smog levels in the eastern U.S. cause 159,000 trips to the emergency room, 53,000 hospital admissions, and 6 million asthma attacks each summer.

Soot and Acid Rain

The burning of coal emits sulfur dioxide (SO_2) and nitrogen oxide (NO_x) gases, which can form fine particles, or soot, when they react with the atmosphere. In addition, coal-fired power plants also emit soot directly from their smokestacks. Scientists increasingly believe soot to be the most dangerous air pollutant, blaming it for 64,000 deaths per year in the U.S., which is almost twice the number of deaths due to auto crashes. Cutting power pl by 75% would avoid more than 18,000 of those deaths.

Soot causes bacterial and viral respiratory infections like pneun as chronic lung diseases, like asthma, that destroy lives over t years. Soot from power plants triggers an estimated 603,000 asthma attacks nationwide every year. Bringing old plants up to modern standards would avoid 366,000 of these attacks. In addition, studies have found that soot may cause heart attacks and arrhythmia (irregular heartbeat) and that the incidence of strokes and heart failure is greater in areas with high levels of soot.

> ***"Even our national parks have not escaped the smog caused by coal-fired power plants."***

Acid rain is formed when sulfur dioxide (SO_2) and nitrogen oxide (NO_x) react with water and oxygen in the atmosphere to form acidic compounds, most commonly sulfuric and nitric acid. These compounds can become incorporated into natural precipitation and fall to the earth as rain or snow. Coal-fired power plants are the largest source of SO_2, 66%, and [are] second to automobiles in NO_x emissions. The Northeast and eastern Canada are home to some of the worst acid rain pollution, because emissions produced from large dirty midwestern coal power plants waft in the wind toward the northeast. For instance, numerous lakes and streams in the Adirondack mountains of upstate New York are too acidic to support fish life, and half of Virginia's native trout streams have reduced capacity due to acid rain.

Acid rain destroys the ecosystems, including streams and lakes, upsetting the delicate balance and making them unable to support life. It also can destroy forests, killing plant and animal life, and eats away at man-made monuments and buildings, effectively destroying our natural and historical treasures.

While the 1990 amendments to the Clean Air Act have made great progress in reducing SO_2 emissions from many of the midwestern coal power plants, more needs to be done. Too many of the lakes and streams in our country continue to suffer from the devastating effects of acid rain.

Toxins

Power plants are one of the largest sources of toxic metal compound pollution. Together they released more than one billion pounds of toxic pollution in 1998, including 9 million pounds of toxic metals and metal compounds and 750 million pounds of dangerous acid gases. Many of these compounds are known or suspected carcinogens and neurotoxins and can cause acute respiratory problems and aggravate asthma and emphysema.

One of the most dangerous toxins emitted is mercury. Coal contains trace amounts of mercury that are released into the air when the fuel is burned to produce electricity. The health hazard results when mercury falls to the earth with rain, snow, and in dry particles.

ercury causes brain, lung, and kidney damage, as well as reproductive probems, and even death in humans and other animals. Mercury is found in fish after it falls into a lake or stream. Just one drop of mercury can contaminate a 25-acre lake to the point where fish are unsafe to eat, making mercury contamination the most common reason for fish advisories issued by States and Native American tribes. The Environmental Protection Agency (EPA) estimates that at least six million women of childbearing age have levels of mercury in their bodies that exceed what the EPA considers acceptable and that 375,000 babies born each year are at risk of neurological problems due to exposure to mercury in the womb.

Global Warming

Burning fossil fuels such as coal releases carbon dioxide (CO_2) pollution. The U.S. has four percent of the world's population yet emits 25% of the global warming pollution. Power plants emit 40% of U.S. carbon dioxide pollution, the primary global warming pollutant. In 1999, coal-fired power plants alone released 490.5 million metric tons of CO_2 into the atmosphere (32% of the total CO_2 emissions for 1999). Currently there is 30% more CO_2 in the atmosphere than there was at the start of the Industrial Revolution, and we are well on the way to doubling CO_2 levels in the atmosphere during this century.

The 1990s were the hottest decade on record. Average global temperatures rose one degree Fahrenheit during the last century and the latest projections are for an average temperature increase of two to as much as ten degrees during this century. In February 2001, the Intergovernmental Panel on Climate Change (IPCC) reported that global warming threatens human populations and the world's ecosystems with worsening heat waves, floods, drought, extreme weather and by spreading infectious diseases. To address the problem of global warming, steps need to be taken to slash the amount of CO_2 power plants emit. We need to switch from burning coal to cleaner burning natural gas and dramatically increase energy efficiency and renewable wind and solar energy.

> ***"Cutting power plant pollutants by 75% would avoid more than 18,000 [soot-related] deaths."***

What Can Be Done?

The government should expand the Clean Air Act to include protections from old and dirty power plants and provide incentives for the use of cleaner fuels. The government should also work towards the replacement of the existing infrastructure with a more sustainable means of producing electricity.

Congress is considering the Clean Power Act and Clean Smokestacks Act of 2001. These companion bills would dramatically cut power plant emissions for four major pollutants by 2007. Smog-forming nitrogen oxide emissions would be cut by 75%, acid rain-forming sulfur dioxide would also be cut by 75%, toxic mercury emissions would be cut by 90%, and carbon dioxide emissions would return to 1990 levels. In addition, these companion bills would require every power plant to meet the most recent pollution controls required for new sources.

Individuals can help by conserving electricity in the home and office by:

- replacing incandescent light bulbs with compact fluorescent bulbs;
- caulking and weather-strip doorways and windows;
- installing low-flow showerheads and faucets;
- keeping the furnace and air conditioner working properly;
- buying energy-efficient electronics and appliances and make sure to turn them off when they're not in use;
- raise awareness in the community by speaking with friends and neighbors and by writing, faxing, calling and emailing to representatives in government and the President.

Chlorofluorocarbons Are Destroying the Ozone Layer

by Linda Baker

About the author: *Linda Baker is a freelance writer concerned with environmental issues.*

In early August, Bert Ammons of Stuart, Florida pleaded guilty to violating the Clean Air Act when he attempted to smuggle ninety 30-pound cylinders of CFC-12, also known by its trade name, Freon, in false compartments on his 41-foot boat, Sierra. According to Environmental Protection Agency (EPA) officials, not only did Ammons plan to distribute the chlorofluorocarbons (CFCs) to auto repair shops around Fort Lauderdale, but his ozone-depleting cargo also had a street value of approximately $68,000. If convicted, Ammons faces up to five years in prison and $250,000 in fines.

Smuggling of CFCs Imperils Ozone Layer

With millions of dollars in evaded taxes and illegal contraband, not to mention multi-agency federal initiatives with names like "Operation Cool Breeze," it's a wonder no one has written a Hollywood thriller about refrigerant and fire-suppressant smugglers. But what illegal CFCs lack in cultural cachet, they make up for in volume and profitability. Between 1994 and 1997, 6,367 tons of CFC-12 and 24 tons of CFC-113 (used as a fire suppressant) were smuggled across the U.S. border. That comes to $43 million in attempted tax evasion alone.

According to an unnamed official in the EPA's Criminal Enforcement Division, illegal CFCs rank close to cocaine as some of the most profitable contraband coming across the U.S. border. The public may not be aware of it, but "black market CFC smuggling is considered a serious problem," says Jack McQuade, an officer with the U.S. Customs Service.

Over the past 10 years, 173 countries including the U.S. have signed the landmark Montreal Protocol of 1987, a global agreement to phase out ozone-depleting chemicals. But the persistent trade in illegal CFCs is only one sign that ozone recovery is far from a sure thing. Recent scientific findings link global warming to ozone depletion, challenging prevailing assumptions that the ozone hole will begin to recover by the year 2050. In October, a major ozone hole opened for the first time over a populated city, Punta Arenas, Chile.

"Policy makers on down say: 'We solved the ozone layer problem. What's next?'" says Kert Davies, science policy director at Ozone Action, a Washington D.C.–based public interest group. "We did the easy thing: We got rid of the CFCs. But when you try to get people to talk about methyl bromide and ozone depletion, about global warming and ozone depletion, it's like pulling teeth."

Declaring Victory?

No one disputes that stratospheric ozone recovery is one of the environmental movement's great success stories. In the 1970s, scientists first discovered that CFCs and other chemicals could damage the Earth's protective ozone layer, which shields life on Earth from the harmful effects of ultraviolet (UV) radiation. These concerns were substantiated in the 1980s by the discovery of the "ozone hole," a thinning of the ozone layer over Antarctica. Additional studies showed that ozone depletion and the corresponding increase in UV radiation hitting the Earth's surface, can have serious consequences for human health and the environment.

Incorporating science, technology and economics, the Montreal Protocol laid out timetables for every country to phase out production of CFCs. In the U.S., Congress amended the Clean Air Act to comply with treaty goals. The scientific community was also charged with re-evaluating the treaty and making amendments accordingly. In 1987, for example, the Protocol called only for a partial phase-out of ozone-depleting chemicals. But re-evaluations in 1989 resulted in a total phase-out of CFCs. Additional assessments in the 1990s led to a dramatic acceleration of the phase-out of the new chemicals, hydrochlorofluorocarbons (HCFCs) and methyl bromide.

"Illegal CFCs [chlorofluorocarbons] rank close to cocaine as some of the most profitable contraband coming across the U.S. border."

"The ozone depletion issue is a good example of the no-net-harm principle combined with the precautionary principle—acting on our knowledge when we have the presumption of a problem," says Davies. "We discovered this hole, we thought there was a link to CFCs and we started moving." The time it takes to get the ball rolling on an international treaty is so great, says Davies, that by the time the Protocol was in place, scientists knew even more about ozone depletion and were able to act accordingly.

At the beginning of the 21st century, experts agree that a world without the

Montreal Protocol would be a horrendous one indeed. According to the Protocol's latest scientific assessment, the world in 2050, absent the global agreement, would look like this: Ozone depletion would be at least 50 percent at mid latitudes. Surface ultraviolet radiation would double at mid latitudes in the Northern Hemisphere and quadruple at mid latitudes in the Southern Hemisphere. By the year 2060, there would be 19 million more cases of nonmelanoma skin cancer and 1.5 million more cases of melanoma skin cancer. And then there would be the numerous unquantifiable effects, such as loss of immunity, lower productivity of crops and damage to aquatic ecosystems.

"The kind of global disaster we averted . . . is indescribable," says John Passacantando, the former executive director of Ozone Action, now head of [environmental activist group] Greenpeace USA. "Had we not phased out this stuff, there would be so much chlorine in the stratosphere it would be like the scene of a bad movie."

The Threat Continues

But temporarily thwarting apocalypse, experts caution, is no cause for complacency. According to the World Meteorological Organization (WMO), in September 1998, the ozone hole over Antarctica was larger and deeper than ever before measured; at 27 million square miles, it covered a surface area larger than North America. The ozone hole over the Arctic also deepened this year, with potentially far more damaging effects on human health.

"In September 1998, the ozone hole over Antarctica was larger and deeper than ever before measured."

The continued threat to the ozone layer can be explained in both political and scientific terms. Under the Montreal Protocol, developing countries have delayed timetables for ending production of ozone-depleting chemicals. Although industrialized countries were required to phase out CFCs by 1996 and methyl bromide by 2005, developing countries have until 2010 to phase out CFCs and until 2015 to phase out methyl bromide.

"The consumption of these chemicals in developing countries is still somewhere around 200,000 tons," says Dr. Omar El-Arini, chief officer of the United Nations Multilateral Fund Secretariat in Montreal, which was established in 1990 to help developing countries comply with terms of the Montreal Protocol. The ozone layer will not recover without the participation of developing nations, he says. "There is only one sky and one ozone layer, which cannot be partitioned."

Psst, Wanna Buy Some . . . Ozone?

Here's where the flourishing trade in illegal refrigerant comes in: CFC production not only continues in developing countries, it is dirt cheap to buy. Sources in the EPA's criminal enforcement division say that in Mexico and

China (among other developing nations), CFC-12 can be bought for $1 or $2 a pound and resold in the U.S. for $20 or $25 a pound. Why the huge domestic markup? It's a simple matter of supply and demand.

With some exceptions for medical use, and use in space shuttle equipment, the United States banned the import of CFCs and other ozone-depleting chemicals in 1996. However, millions of pieces of equipment that use CFCs are still in service, including most automobiles built before 1994, air conditioners and other refrigeration equipment.

"Until older cars get off the road, and until developing countries stop producing CFCs, . . . CFC smuggling will continue."

Although it's possible to retrofit much of this equipment to be serviced with ozone-friendly alternatives, costs can run anywhere from a few hundred to thousands of dollars. Contributing to the problem, the U.S. and other industrialized countries allow the trade and use of recycled CFCs to maintain existing machinery. Because it's almost impossible to distinguish between new and recycled Freon, traders illegally bring CFCs into industrialized countries in the guise of recycled substances or exports to developing countries. A high excise tax on the sale or use of CFCs in the U.S. ($5.35 per pound) also abets the illegal trade.

EPA and Customs Service officers say it's impossible to estimate the quantity of illegal CFCs crossing the border. Nonetheless, the scope of the black market is startling. From Russia to Australia, federal officials paint a picture of worldwide CFC smuggling operations that run the gamut from small-time entrepreneurs to sophisticated money laundering conspiracies.

Since the launch of a nationwide CFC enforcement initiative in 1995, which involves the United States Customs Service, the EPA, the FBI and the IRS, over 100 individuals have been convicted for violation of customs law and the federal Clean Air Act. Defendants included Richard Schmolke, who was convicted last year for a scheme to illegally import 75,000 pounds of CFC-12 from Venezuela into Texas. Agents said that Schmolke was part of one of the largest Freon smuggling rings they had ever encountered.

Federal officials anticipate an increase in smuggling activity as the supply of legal CFCs is depleted . . . said Jack McQuade of the U.S. Customs Service. As of July 2000, a total of 5,438 pounds of CFC-12 had been seized along the southern border and 2,700 pounds of CFC-12 had been taken in the south Florida area, he says.

The reports sound like a parody of film noir. During summer 2000, the U.S. Customs Service regularly intercepted "frio banditos" coming across the Rio Grande, all with cylinders of CFCs strapped to their backs. Geographically, this is the frontline for new smuggling rings. "There are indications that consolidation of individually smuggled CFCs is now occurring along the U.S./Mexican border," says McQuade.

Until older cars get off the road, and until developing countries stop producing CFCs, say EPA officials, CFC smuggling will continue. A related problem, according to El-Arini, is that industrialized countries are dumping CFC-containing products and equipment on developing nations. This will further complicate efforts by developing countries to comply with the Montreal Protocol, he says.

Despite widespread "cheating" and smuggling, the global effort to restore the ozone layer is a remarkable achievement, especially when viewed from an international political perspective. The Multilateral Fund, for example, helped the majority of developing countries freeze production and consumption of ozone-depleting substances at 1999 levels—the first Protocol obligation for these countries. Since 1991, the Fund has disbursed more than $1 billion to phase out the consumption of 142,000 tons of ozone-depleting chemicals in over 110 developing countries.

"This is the first real-life endeavor of mankind to join hands to solve environmental damage that threatens our common habitat, the Earth," says El-Arini. "It proves that once the political commitment is there, national borders can be crossed to overcome a problem of a global dimension."

Trouble at Home: Methyl Bromide Use Continues

Back at home, those familiar with the political scene aren't quite so sanguine. This year, the U.S. is headed for a showdown over methyl bromide, a toxic ozone-depleting chemical used in this country primarily by California strawberry and Florida tomato growers. Introduced last spring by California Congressman Richard Pombo, the un-ironically titled Methyl Bromide Fairness Act would push back the U.S. phase-out date to 2015—the year developing countries are required to stop production and consumption of the chemical.

Due to its acute toxicity, methyl bromide is already banned in several countries, including the Netherlands and Canada. For years, environmentalists and health officials in the U.S. (which uses 40 percent of the world's methyl bromide) have called for stricter regulation of the pesticide, especially in agricultural areas such as California's Ventura County, where children and farm workers are at risk. Since 1982, nearly 500 poisonings linked to methyl bromide have occurred in California, 19 of them fatal.

"Since 1982, nearly 500 poisonings linked to methyl bromide have occurred in California, 19 of them fatal."

The best account of methyl bromide's tarnished history in American politics can be found in a report published by the Transnational Resource and Action Center (TRAC) and the Political Ecology Group (PEG) in 1997. Titled "Bromide Barons: Methyl Bromide, Corporate Power and Environmental Justice" the report meticulously documents how the Big Three methyl bromide corporations, Albemarle, Great Lakes

Chemical and Dead Sea Bromine, as well as California-based TriCal, the largest applicator of methyl bromide in the state, have systematically worked to roll back environmental regulations that threaten profit margins.

"Through various industry groups," the report states, "including the Methyl Bromide Global Coalition and the Methyl Bromide Working Group, the bromide barons have hindered the development of alternatives to methyl bromide, cast doubt on the scientific consensus that methyl bromide contributes to ozone depletion, and influenced the political process through lobbying."

"Recent discoveries about ozone depletion . . . are forcing scientists to revise earlier claims that the ozone layer will begin to recover by 2050."

In 1998, industry-backed congressional representatives tried—and failed—to pass a bill that would push back the phase-out date of methyl bromide. With an environmental rider to the 1999 budget, they succeeded: the date was bumped four years to 2005. Now it's round three. At a House Agriculture Subcommittee hearing July 2000 on the Methyl Bromide Fairness Act (which now has more than 20 sponsors), efforts to undermine the methyl bromide ban were relentless.

"While methyl bromide has been placed in the position of public enemy number one by the radical environmental community, we have lost sight of the fact that this may truly be a silver bullet compound," excoriated Jim Culbertson, executive manager of the California Cherry Export Association. "The sky is not falling and agricultural methyl bromide is not the cause of the ozone hole."

Claims that methyl bromide has a negligible effect on ozone depletion are simply not true, counters Azadeh Tabazadeh, an atmospheric chemist at the NASA Ames Research Center in Mountain View, California. "In fact, the bromine in methyl bromide is a much better catalyst for ozone destruction than chlorine," she says. "And just because we've reduced the amount of chlorine in the atmosphere doesn't mean that the level of bromine is also going down. That's why compounds like methyl bromide need to be regulated."

The government and scientific community agree. The EPA identifies methyl bromide as a Class I ozone-depleting substance that will be phased out under the Clean Air Act. A 1994 paper, co-authored by several federal agencies and the United Nations Environment Programme (UNEP), warns that with the phase-out of CFCs underway, the elimination of methyl bromide emissions "from agricultural, structural, and industrial activities" is the single most important step that the world's governments can take to reduce future levels of ozone depletion.

Case studies listed by the EPA demonstrate that viable chemical and organic substitutes for methyl bromide do exist. "Farmers are reluctant to change because their crop production systems have been developed around methyl bromide," says Bill Thomas of the EPA's Stratospheric Protection Division. "But a number of good alternatives to methyl bromide are now available, which should allow most growers to continue to produce their crops in a way they're used to."

Feeding the Loop: Global Warming and Ozone Depletion

As U.S. special interests backpedal on methyl bromide, recent discoveries about ozone depletion in the Northern Hemisphere are forcing scientists to revise earlier claims that the ozone layer will begin to recover by 2050. Although satellites have detected an ozone "cavity" over the Arctic for several years, the phenomenon is growing worse. From November 1999 through March 2000, seasonal ozone concentrations in some parts of the Arctic declined as much as 60 percent.

From a human health perspective, Arctic ozone thinning is more worrisome than comparable reductions over Antarctica. This is because ozone-depleted air from the Arctic drifts south each spring toward highly populated areas in North America, Europe and Russia. Last year, a European Space Agency's satellite revealed that ozone levels in Great Britain, Belgium, Netherlands and Scandinavia were nearly as low as those normally found in the Antarctic.

There's another reason why the ozone hole over the Arctic is attracting attention. Scientists have known for some time that ozone lows are often associated with extremely low temperatures in the stratosphere and the presence of polar stratospheric clouds, which provide the template for the chemical reactions that destroy ozone. Polar clouds are also a common presence in the Antarctic, where temperatures are colder than the Arctic.

"The fight to protect the ozone layer has become a model for global environmental protection."

Here's the key finding: Over the past couple of decades, the Arctic has become more like the Antarctic; in other words, it's getting a lot colder. And as recent studies published in the journal *Science* suggest, global warming may be the culprit.

"It's ironic because people have always been confused about ozone depletion and the greenhouse effect—the general public always thought they were intertwined," says Davies. "And now it turns out they are."

The feedback loop between global warming and ozone depletion works like this: The warming of the lower atmosphere known as the greenhouse effect traps warm air at the surface. This in turn leads to cooling in the upper atmosphere, which creates the conditions for ozone depletion to take place. CFCs, which deplete ozone, are also a culprit in global warming.

Scientists used to believe that as chlorine levels declined in the upper atmosphere, the ozone layer would start to recover, says Tabazadeh, who co-authored a recent study on the role polar clouds assume in ozone depletion. "That would be true only if the climate was persistently the same," she says. "But if the climate is getting colder due to surface warming, the upper atmosphere is primed for massive destruction of ozone. Things are going to get worse before they get better."

The discovery highlights yet another set of economic, environmental and political problems—namely, what to do about hydrofluorocarbons (HFCs). HFCs were originally introduced as an ozone-friendly alternative to CFCs; however, they are now recognized as a powerful greenhouse gas, with as much as 4,000 times the global warming potential of carbon dioxide.

So far, the HFC issue has underscored tensions between groups concerned about global warming and groups working toward ozone recovery. For example, this year Coca-Cola announced plans to phase out the use of HFCs in its cold-drink equipment. The move was applauded by environmental groups like Greenpeace but criticized by industry groups such as the Alliance for Responsible Atmospheric Policy, which said it would threaten the ozone layer as well as the economic competitiveness of companies that have invested millions in HFC technology.

The fight to protect the ozone layer has become a model for global environmental protection. But as the continuing battle over methyl bromide, the illegal trade in CFCs and now the controversy over HFCs suggest, environmental memory is not only short term—it can also be short-circuited. Whether new scientific discoveries result in more holistic public policies remains to be seen.

"The atmosphere has the potential to be the big wake-up call on the environment," says Davies. "Because the more we look, the more we see that all these issues are connected. The atmosphere is the ultimate global commons."

Automobiles Cause Air Pollution

by the American Council for an Energy-Efficient Economy

About the author: *The American Council for an Energy-Efficient Economy (ACEEE) is an organization working to promote economic prosperity and environmental protection through the advancement of energy efficiency.*

Automobiles affect the environment in many ways. Impacts begin when a vehicle is manufactured (including the production of all the parts and materials that go into the car) and end with its scrappage in a junkyard (which can recycle many parts but also involves the disposal of many wastes). Over the life of an average motor vehicle, however, much of the environmental damage occurs during driving and is greatly associated with fuel consumption. . . . Nearly 90 percent [of lifecycle energy use for a typical automobile] is due to fuel consumption over the dozen or so years of a vehicle's life.

Automobiles and the Environment

Environmental impacts start with mineral extraction and the production of the raw materials that go into the parts of a car. For example, iron ore gets turned into steel, which now accounts for most of the mass in vehicles. Steel can be recycled, of course. On average, today's automobiles are about 75 percent recyclable, and using recycled steel helps reduce energy use and pollution. Other metal components, such as aluminum (used in some engine parts and wheels, for example) and copper (used for wiring) are also largely recycled. The lead and acid in batteries are poisonous and dangerous. But batteries can be recycled, if they are returned to a service station, a parts store, or brought to a municipal hazardous waste facility. Plastics, which are mostly made from petroleum, are more difficult to recycle. In any case, some degree of pollution is associated with all of these components, much of it due to the energy consumption, air pollution, and releases of toxic substances that occur when automobiles are manufactured and distributed.

Most of the environmental impact associated with motor vehicles occurs when they are used, due to pollution in their exhaust and pollution associated with supplying the fuel. In the United States, nearly all of today's automobiles use gasoline; a lesser number use diesel fuel. In some areas, various alternative fuels are being introduced, but these are not widely available for most drivers. When gasoline, diesel, or other fuels are burned in car engines, combustion is never perfect, and so a mix of hazardous pollutants comes out the tailpipe.

If combustion were perfect and didn't create noxious by-products, the exhaust would contain only water vapor and carbon dioxide. Carbon dioxide (CO_2) isn't directly harmful to health, at least not in low concentrations. After all, CO_2 is also what we exhale as a result of our "burning" the calories in the food we eat. However, CO_2 from fossil fuels like gasoline and diesel is very harmful to the environment because it causes global warming—more on this pollutant shortly.

Motor fuel is itself a product and so, like a car, environmental damage occurs throughout its lifecycle as well. For gasoline and diesel, the product lifecycle begins at the oil well and ends when the fuel is burned in the engine. *Fuel cycle* impacts are the forms of pollution and other environmental damage that occur between the oil well and the fuel tank. Gasoline and diesel fuel are poisonous to humans, plants, and animals, and their vapors are toxic. Other energy sources have their own fuel cycles. With battery-powered electric vehicles, for example, no fuel is burned onboard the vehicle, and so nearly all of the fuel-cycle pollution and energy use occurs at electric power plants and in producing the fuels that run the power plants. Many of the same air pollutants that spew from vehicle tailpipes are also spewed from power plants and oil refineries (as well as the tanker trucks that deliver gasoline to your local filling station).

A Harmful Addiction to Oil

Gasoline and diesel fuel now provide 97 percent of America's transportation energy needs. Air pollution isn't the only problem associated with these petroleum-based fuels. Oil extraction lays waste to many fragile ecosystems, harming tropical forests in South America and Southeast Asia, deserts and wetlands in the Middle East, our own coastal areas, and the fragile tundra and arctic coastal plains of Alaska. Millions of gallons of oil are spilled every year. Sometimes the disasters are well known, such as the 1989 Exxon Valdez spill in Prince William Sound. More often there are rarely reported but still tragic smaller spills that occur in the oceans and in coastal waters, bays, and rivers throughout the world. In our own communities, groundwater is sometimes tainted by leaks from underground fuel storage tanks and mis-

"When gasoline, diesel, or other fuels are burned in car engines . . . a mix of hazardous pollutants comes out the tailpipe."

cellaneous spills that occur during shipping and handling of the 120 billion gallons of fuel we use each year.

In addition to these environmental harms, gasoline and diesel consumption bring economic and security risks. The Middle East contains the largest concentration of the world's oil. The United States maintains a global military presence partly to maintain access to foreign oil. Most recently in the 1991 war with Iraq, but several times throughout our history, U.S. troops were committed and blood was spilled to secure our oil supply.

"The average light truck pollutes more than the average car."

Major recessions were triggered by oil crises in the 1970s and early 1980s, causing unemployment and inflation. Oil imports drain over $50 billion per year from American pockets, representing lost job opportunities even when our economy seems to be doing fine. Half of U.S. oil is now imported and our dependence on foreign sources is steadily rising, perpetuating the risk of future oil crises. The past year's run-up in gasoline prices is just the latest example of how petroleum dependence can squeeze family budgets only to enrich oil producers.

Our addiction to gasoline and diesel fuel also involves moral compromises. It entails deals and economic arrangements with some oil-rich countries whose standards of human rights and environmental protection may not be the same as what we expect at home. Of course, these issues go beyond strictly environmental concerns. Nevertheless, choosing greener vehicles that consume less fuel not only protects the environment, but also helps protect U.S. jobs while reducing the security risks and moral liabilities of oil dependence.

Major Pollutants Associated with Automobiles

Our focus . . . is on air pollutants related to car and truck fuel consumption, because they are such a large part of a vehicle's environmental damage and because they are the main impacts that can be reduced through your choice of make and model. . . . The pollution coming from vehicles can differ depending on the standards they meet (and how well their emissions controls work), how they are driven and maintained, and the quantity and quality of the fuel they burn. Many vans, pickups, sport utilities, and other light trucks meet less stringent emissions and fuel economy standards than vehicles classified as passenger cars. As a result . . . the average light truck pollutes more than the average car.

All new vehicles must meet either the emissions standards set by the U.S. Environmental Protection Agency (EPA) or those set by the California Air Resources Board (CARB). Generally, California standards are more stringent than the Federal standards. A number of Northeastern states have adopted the California standards, and vehicles meeting these standards are becoming more and more common nationwide.

Vehicles are also tested for fuel economy, as measured by miles per gallon—

MPG. Fuel economy standards apply to manufacturers, rather than individual vehicles, and are set so that an automaker can sell a variety of vehicles as long as the average MPG of its sales meets the applicable standard. Manufacturers calculate the fuel economy of each model they sell using laboratory tests similar to those conducted to determine tailpipe emissions. Because these tests give fuel economy estimates higher than what most people experience in real-world driving, the MPG measurements are adjusted downward by EPA. These adjusted MPG numbers are printed on new vehicle stickers. . . .

Although a wide variety of pollutants are formed in the various stages of an automobile's lifecycle, our ratings are mainly based on the serious air pollutants that are regulated to control vehicle emissions. All of these pollutants are more damaging to health when emitted from vehicle tailpipes than when a similar quantity is emitted from a power plant, since tailpipe pollution is literally "in your face," subjecting people to more direct exposures during daily activities.

Particulate Matter (PM)

Fine airborne particles are an established cause of lung problems, from shortness of breath to worsening of respiratory and cardiovascular disease, damage to lung tissues, and cancer. Certain people are particularly vulnerable to breathing air polluted by fine particles, among them asthmatics, individuals with the flu and with chronic heart or lung diseases, as well as children and the elderly. PM also soils and damages buildings and materials. It forms haze that obscures visibility in many regions. Soot and smoke coming from exhaust pipes are obvious sources of PM, but among the most deadly forms of airborne particulate matter are the invisible fine particles that lodge deeply in the lungs. PM has been regulated for some time, but the regulations were based on counting all particles up to 10 microns in size (PM10). However, PM10 standards fail to adequately control the most dangerous, very fine particles. The U.S. EPA has recently started to regulate fine particles up to 2.5 microns in size (PM2.5), which better focuses on the most damaging category.

Properly functioning new, fuel-injected gasoline vehicles directly emit very little PM2.5. But they indirectly cause significant PM pollution as a result of their NO_x, SO_2, and HC emissions, not only from tailpipe but also from vehicle manufacturing and fuel refining. These emissions result in "secondary" particle formation. This phenomenon refers to the way that the gaseous pollutants agglomerate ("glom up") at microscopic scales to form fine particles that are largely invisible but cause the health problems mentioned. Transportation sources account for about 20 percent of directly emitted PM2.5. Diesel engines are the major source of direct PM emissions from motor vehicles. Although most such emissions come from

> ***"Gasoline vehicles . . . indirectly cause significant PM [particulate matter] pollution as a result of their . . . emissions."***

heavy trucks and diesel buses, even the smaller diesel engines in some cars and light trucks emit significant amounts of fine PM.

Nitrogen Oxides (NO_x)

NO_x refers mainly to two chemicals, nitrogen oxide (NO) and nitrogen dioxide (NO_2) that are formed when nitrogen gas, which comprises 78 percent of air, reacts with oxygen during the high temperatures that occur during fuel combustion. NO_x is truly a noxious pollutant in many ways. It is directly hazardous, an irritant to the lungs that can aggravate respiratory problems. It reacts with organic compounds in the air to cause ozone, which is the main reason for "smog alerts" that still happen too often in many cities and regions. NO_x is a precursor of fine particles, which cause respiratory problems and lead to thousands of premature deaths each year. It is also a precursor of acid rain, which harms lakes, waterways, forests, and other ecosystems, as well as damaging buildings and crops. Airborne NO_x also contributes to nitrification—essentially an over-fertilization of wetlands and bays, leading to algae blooms and fish kills.

As an air pollutant, NO_x is one of the most difficult to control since it is such a pervasive product of combustion. Nationwide, most NO_x comes from electric power plants and industrial sources. Natural gas and oil-fired home furnaces and water heaters also produce NO_x in their flue gases. Motor vehicles account for about one-third of nationwide NO_x emissions. Many of these emissions come from heavy-duty diesel trucks, but cars and light trucks are also a major source. NO_x has also been one of the most difficult pollutants to get out of our air. EPA air quality regulations have helped keep emissions from growing as fast as they might have, but nationwide, overall NO_x emissions are higher than they were a decade ago. A good portion of this growth in NO_x pollution has been from cars and light trucks.

"Cleaner, low sulfur fuels . . . would greatly reduce . . . pollution from all cars and trucks on the road."

Sulfur Dioxide (SO_2)

Gasoline and diesel fuels also contain varying amounts of sulfur, which burns in the engine to produce sulfur dioxide (SO_2). This gaseous chemical is another source of secondary particulate formation, and is itself a lung irritant as well as a cause of acid rain. SO_2 also interferes with the operation of catalytic converters. Some of the cleaner, reformulated versions of gasoline have very low sulfur levels. Most gasoline sold nationwide still has too much sulfur, but levels are being reduced under recently established EPA regulations.

Cars and light trucks are not the largest source of SO_2 emissions, which come mainly from power plants and industrial facilities. However, because cars and light trucks are so numerous and gasoline has a high average sulfur content, cars and light trucks cause twice as much fine PM pollution as heavy freight trucks. Making all gasoline as clean as the cleaner, low-sulfur fuels already available in California would greatly reduce this PM pollution from all cars and trucks on the road, both new and used.

Hydrocarbons (HC)

Hydrocarbons are a broad class of chemicals containing carbon and hydrogen. Those hydrocarbons that cause various forms of air pollution are also known as volatile organic compounds since they are forms of HC that are either gases or readily evaporate into the air. Many forms of HC are directly hazardous, contributing to what are collectively called "air toxics." These compounds can be directly irritating to the lungs and other tissues and they can also cause cancer, contribute to birth defects, and cause other illnesses. During daylight hours, and particularly during hot summer weather, HC reacts with NO_x to form ozone smog. Controlling ozone is one of the major environmental challenges in the United States. Although progress has been made over the past several decades, many cities and regions still have smog alerts when ozone levels get too high.

Gasoline vapor contains a mix of hydrocarbons. Thus, HC pollution is produced whenever we fill our tanks. Some regions have special nozzles on fuel pumps to help trap such vapors. Other HC vapors are released at various stages along the way from the refinery to the filling station. Vapors seep out, even when a car is parked and turned off, due to the imperfect sealing of the fuel tank, pipes and hoses, and other components leading to the engine. HC also comes out of the tailpipe, as a result of incomplete combustion and the less-than-perfect cleanup of exhaust gases by catalytic converters and other vehicle emissions controls. Diesel fuel is less volatile than gasoline, so evaporation is less of a problem. Nevertheless, diesel exhaust still contains many toxic hydrocarbons and other compounds. Overall, transportation is responsible for about 36 percent of man-made HC emissions in the United States.

Toxic Chemicals

Toxic releases are just that—any number of a wide range of chemicals that can cause cancer, birth defects, cardiovascular, respiratory, and neurological damage, or other forms of health harm. Many smog-forming hydrocarbons are directly toxic; for example, benzene is a known human carcinogen. Other toxics include solvents and metallic compounds such as lead and chromium. Toxics are released during many industrial activities and car and truck manufacturing is a significant source. Workers and communities near factories are at the highest risk. Vehicles also emit toxics, due to fuel evaporation

> ***"Diesel exhaust . . . contains many toxic hydrocarbons and other compounds."***

while pumping gas and while a car sits in the sun, for example, as well as from the tailpipe. Diesel exhaust, in particular, has been implicated as a harmful toxic release.

Toxic emissions from cars and trucks, as well as toxic releases during the production and assembly of vehicles and their components, are controlled by various regulations. Factories and other manufacturing facilities are required to report toxic emissions from each site. But controls are far from perfect, and there are many ways in which industry could do a better job of preventing toxic pollution. . . .

Carbon Monoxide (CO)

Carbon monoxide is an odorless, colorless, but potentially deadly gas that is created by the incomplete combustion of any carbon-containing fuel, including gasoline and diesel. When inhaled, CO combines with the hemoglobin in our blood, impairing the flow of oxygen to our brain and other parts of the body. We've all heard stories of people being killed by carbon monoxide poisoning, from vehicles in closed garages, during fires, or in homes when indoor CO concentrations are raised by malfunctioning stoves or furnaces. Even if it doesn't cause death, CO exposure can cause permanent damage to the nervous system. At lower concentrations, CO is still harmful, particularly for people with heart disease. In some areas, cars and trucks can create enough CO to cause health risks outdoors.

"Motor vehicles are responsible for about 60 percent of CO [carbon monoxide] emissions nationwide."

Large amounts of CO are produced when a vehicle first starts up and its engine is cold. Poorly designed and malfunctioning engines and emission controls systems are also responsible for excess CO pollution. Motor vehicles are responsible for about 60 percent of CO emissions nationwide.

Cars, Trucks, and Global Warming

The gasoline-powered automobile was invented just over 100 years ago, when the industrial revolution was still young. Streams had long been dammed to turn mills, and coal was on its way to widespread use—it was already powering steamships and locomotives. But most energy used by humans still came from traditional fuels such as wood. In 1890, the world population was about 1.5 billion but growing rapidly. The amount of carbon dioxide (CO_2) in the atmosphere was just over 290 parts per million, not yet noticeably over its level throughout pre-industrial civilization.

The world population has now topped six billion and is still growing rapidly. During the past century, the amount of fossil fuel we consume has risen nearly five times faster than population. As a result, the amount of CO_2 in the atmosphere is now over 360 parts per million and climbing. This rapid increase in CO_2 concentration represents an enormous impact of our energy-consumptive lifestyle on the planet, and it is causing dangerous changes to the earth's climate. The past decade has already seen many years with above-normal temperatures. The changes in weather patterns and increases in severe events are consistent with climate disruption. Recent years have been among the warmest ever recorded.

"The increased risks [of global warming] have created a call . . . to curtail CO_2 [carbon dioxide] emissions around the world."

Carbon dioxide is the most important of what are known as greenhouse gases, compounds that enable the earth's atmosphere to trap heat, like a greenhouse, but on a global scale. Too much greenhouse gas in the atmosphere causes global warming, an increase in global average temperatures above what they normally would be.

The risks of global warming are many. Human health is threatened by more frequent and severe heat waves and the spread of tropical diseases. Lives can be lost because of rising sea levels and more severe storms, which can also damage regional and national economies. The disruptions to climate are unpredictable but certainly risky. While some areas may see greater coastal flooding and inundating rains, other regions may experience droughts. Both agriculture and natural habitats can be harmed. Future generations will bear the brunt of these risks, but the effects of global warming have already been detected. Although we cannot attribute any given event to climate change, the increased risks have created a call for action to curtail CO_2 emissions around the world.

Oil is now the world's dominant fuel. There are over 600 million cars and trucks in the world. Both here and abroad, transportation accounts for most oil use. In the United States, we now have more motor vehicles than licensed drivers, and we travel over 2 trillion miles per year, burning 120 billion gallons of gasoline. Not counting the "upstream" emissions from producing the fuel, the result is over a billion tons of CO_2 pollution each year.

U.S. cars and light trucks alone account for more energy-related CO_2 than the nationwide emissions of all but three other countries in the world (China, Russia, and Japan). Our vehicles produce more CO_2 than all of India, which has more than triple our population. U.S. cars and trucks emit twice as much fossil-fuel CO_2 as the economies of either South Korea or Mexico and over three times as much as the whole of Brazil. Although some of these countries are growing and industrializing rapidly, it will be decades before their level of CO_2 pollution per person approaches ours.

Fuel Efficiency and Air Pollution

The amount of CO_2 emitted by a vehicle is essentially proportional to the amount of fuel burned. Thus, fuel-efficient vehicles are the best choice for helping to stop global warming. And gas guzzlers are global polluters.

For other forms of air pollution, the relation between fuel economy and emissions is more complex. Automobile emissions are regulated to a given number of grams per mile, independently of how much fuel they burn (although standards are weaker for many gas-guzzling light trucks). But several factors cause NO_x, HC, CO, and PM pollution to be higher when a vehicle's fuel economy is lower.

In real-world use, most vehicles' emissions are much higher than the standards levels. The reasons include the fact that automakers' and EPA's emissions tests fail to fully represent real-world driving, malfunction of emissions control systems, deterioration of components, inadequate or incorrect maintenance, and sometimes tampering. A portion of this excess pollution is proportional to a vehicle's rate of fuel consumption. Automobiles that meet a more stringent emissions standard are generally cleaner than those that meet a less stringent standard. However, among vehicles that meet the same standard, those with higher fuel economy generally produce less air pollution.

A significant amount of pollution also occurs in supplying vehicles with fuel. These so-called upstream emissions occur everywhere from the oil well and refinery to the filling station and gas tank, before the fuel gets to the engine. The relationship between fuel consumption and upstream emissions is strongest for hydrocarbons (HC). For an average car, about 11 grams of excess HC pollution (beyond what comes out of the tailpipe) occurs for every gallon of gasoline burned. Lesser but still-significant amounts of other pollutants are also related to the amount of fuel burned. Examples include NO_x and PM from tanker trucks delivering gasoline and a whole soup of pollutants from oil refineries. Thus, higher fuel consumption implies higher upstream pollution.

Air Quality Has Improved

by the Environmental Protection Agency

About the author: *The Environmental Protection Agency (EPA) is a department of the federal government that establishes and enforces the nation's environmental laws.*

EPA has set national air quality standards for six principal air pollutants (also referred to as criteria pollutants): carbon monoxide (CO), lead (Pb), nitrogen dioxide (NO_2), ozone (O_3), particulate matter (PM), and sulfur dioxide (SO_2). Four of these pollutants (CO, Pb, NO_2, and SO_2) result solely from direct emissions from a variety of sources. PM can result from direct emissions also, but is commonly formed when emissions of nitrogen oxides (NO_x), SO_2, ammonia, and other gases react in the atmosphere. Ozone is not directly emitted, but is formed when NO_x and volatile organic compounds (VOCs) react in the presence of sunlight.

Each year EPA examines changes in levels of these ambient pollutants and their precursor emissions over time and summarizes the current air pollution status.

Summary of Air Quality and Emissions Trends

EPA tracks trends in *air quality* based on actual measurements of pollutant concentrations in the ambient (outside) air at monitoring sites across the country. Monitoring stations are operated by state, tribal, and local government agencies as well as some federal agencies, including EPA. Trends are derived by averaging direct measurements from these monitoring stations on a yearly basis. The air quality based on concentrations of the principal pollutants has improved nationally over the last 20 years (1981–2000).

EPA estimates nationwide *emissions* of ambient pollutants and their precursors based on actual monitored readings or engineering calculations of the amounts and types of pollutants emitted by vehicles, factories, and other sources. Emission estimates are based on many factors, including the level of industrial activity, technology developments, fuel consumption, vehicle miles traveled, and other activities that cause air pollution. Emissions estimates also reflect changes

Excerpted from *Latest Findings on National Air Quality: 2000 Status and Trends*, by the Environmental Protection Agency, September 2001.

in air pollution regulations and installation of emissions controls. The 2000 emissions reported in this summary report are projected numbers based on available 1999 information and historical trends. EPA's emission estimation methods continue to change and improve. As a result, comparisons of the estimates for a given year in this summary to the same year in previous summaries may not be appropriate. . . . Emissions of the principal pollutants have decreased over the last 20 years (1981–2000), with the exception of NO_x. While NO_x emissions have increased, air quality measurements for NO_2 across the country are below the national air quality standards. It is important to note that oxides of nitrogen, including NO_2, contribute to the formation of ozone, particulate matter, and acid rain. NO_x also add to poor visibility.

"Emissions of the principle pollutants have decreased over the last 20 years."

Between 1970 and 2000, gross domestic product increased 158 percent, energy consumption increased 45 percent, vehicle miles traveled increased 143 percent, and U.S. population increased 36 percent. At the same time, total emissions of the six principal air pollutants decreased 29 percent.

The improvements are a result of effective implementation of clean air laws and regulations, as well as improvements in the efficiency of industrial technologies.

Despite great progress in air quality improvement, approximately 121 million people nationwide still lived in counties with pollution levels above the national air quality standards in 2000.

Nitrogen Dioxide (NO_2)

Nitrogen dioxide (NO_2) is a reddish brown, highly reactive gas that is formed in the ambient air through the oxidation of nitric oxide (NO). Nitrogen oxides (NO_x), the term used to describe the sum of NO, NO_2 and other oxides of nitrogen, play a major role in the formation of ozone, particulate matter, haze and acid rain. The major sources of man-made NO_x emissions are high-temperature combustion processes, such as those occurring in automobiles and power plants. Home heaters and gas stoves also produce substantial amounts of NO_2 in indoor settings. . . .

Over the past 20 years, monitored levels of NO_2 have decreased 14 percent. All areas of the country that once violated the national air quality standard for NO_2 now meet that standard. While levels around urban monitors have fallen, national emissions of nitrogen oxides have actually increased over the past 20 years by 4 percent. This increase is the result of a number of factors, the largest being an increase in nitrogen oxides emissions from diesel vehicles. This increase is of concern because NO_x emissions contribute to the formation of ground-level ozone (smog), but also to other environmental problems, like acid rain and nitrogen loadings to water bodies.

Ground-Level Ozone (O_3)

Ground-level ozone (the primary constituent of smog) continues to be a pollution problem throughout many areas of the United States.

Ozone is not emitted directly into the air but is formed by the reaction of VOCs and NO_x in the presence of heat and sunlight. Ground-level ozone forms readily in the atmosphere, usually during hot summer weather. VOCs are emitted from a variety of sources, including motor vehicles, chemical plants, refineries, factories, consumer and commercial products, and other industrial sources. Nitrogen oxides are emitted from motor vehicles, power plants, and other sources of combustion. Changing weather patterns contribute to yearly differences in ozone concentrations from region to region. Ozone and the precursor pollutants that form ozone also can be transported into an area from pollution sources found hundreds of miles upwind.

Short-term (1–3 hours) and prolonged (6–8 hours) exposures to ambient ozone have been linked to a number of health effects of concern. . . .

In 1997, EPA revised the national ambient air quality standards for ozone by setting new 8-hour 0.08 ppm standards. Currently, EPA is tracking trends based on both the 1-hour and 8-hour data.

> ***"National ambient ozone levels decreased 21 percent."***

Over the past 20 years, national ambient ozone levels decreased 21 percent based on 1-hour data, and 10 percent based on 8-hour data. Between 1981 and 2000, emissions of VOCs have decreased 32 percent. During that same time period, emissions of NO_x increased 4 percent.

Because sunlight and heat play a major role in ozone formation, changing weather patterns contribute to yearly differences in ozone concentrations. To better reflect the changes that emissions have on measured air quality concentrations, EPA is able to make analytical adjustments to account for this annual variability in meteorology. For 52 metropolitan areas, the adjusted trend for 1-hour ozone levels shows improvement over the 20-year period from 1981–2000. However, beginning in 1994, the rate of improvement appears to level off and the trend in the last 10 years is relatively flat.

Regional Trends in Ozone Levels

For the period 1981–2000, the downward trend in 1-hour ozone levels seen nationally is reflected in every broad geographic area in the country. The Northeast and West exhibit the most substantial improvement while the South and Southwest have experienced the least rapid progress in lowering ozone concentrations. Over the last 10 years, this downward trend continues for the Northeast, Midwest and West coast; however, in the South and North Central regions of the country, ozone levels have actually increased.

Across the country, the highest ambient ozone concentrations are typically found at suburban sites, consistent with the downwind transport of emissions from urban centers. During the past 20 years, ozone concentrations decreased more than 24 percent at urban sites and declined by 21 percent at suburban sites. For the more recent 10-year period, urban sites show decreases of approximately 12 percent and suburban sites show 11 percent decreases. However, at rural monitoring locations national improvements have been slower. One-hour ozone levels for 2000 are 15 percent lower than those in 1981 but only 6 percent below 1991 levels. In 2000, for the third consecutive year, rural 1-hour ozone levels are greater than the levels observed for the urban sites, but they are still lower than levels observed at suburban sites.

Over the last 10 years, 8-hour ozone levels in 29 of our national parks increased over 4 percent. Thirteen monitoring sites in eleven of these parks experienced statistically significant upward trends in 8-hour ozone levels: Great Smoky Mountains (TN), Cape Romain (SC), Cowpens (SC), Congaree Swamp (SC), Everglades (FL), Mammoth Cave (KY), Voyageurs (MN), Yellowstone (WY), Yosemite (CA), Canyonlands (UT) and Craters of the Moon (ID). For the remaining 18 parks, the 8-hour ozone levels at ten increased only slightly between 1991 and 2000, while seven showed decreasing levels, and one was unchanged.

Sulfur Dioxide (SO_2)

Sulfur dioxide belongs to the family of sulfur oxide gases. These gases are formed when fuel containing sulfur (mainly coal and oil) is burned and during metal smelting and other industrial processes. Most SO_2 monitoring stations are located in urban areas. The highest monitored concentrations of SO_2 are recorded in the vicinity of large industrial facilities. Fuel combustion, largely from coal-fired power plants, accounts for most of the total SO_2 emissions. . . .

Nationally, average SO_2 ambient concentrations have decreased 50 percent from 1981–2000 and 37 percent over the more recent 10-year period 1991–2000. SO_2 emissions decreased 31 percent from 1981 to 2000 and 24 percent from 1991–2000. Reductions in SO_2 concentrations and emissions since 1994 are due, in large part, to controls implemented under EPA's Acid Rain Program beginning in 1995.

Particulate Matter

Particulate matter (PM) is the general term used for a mixture of solid particles and liquid droplets found in the air. Some particles are large or dark enough to be seen as soot or smoke. Others are so small they can be detected only with an electron microscope. $PM_{2.5}$ describes the "fine" particles that are less than or equal to 2.5 micrometers in diameter. "Coarse" particles are greater than 2.5, but less than or equal to 10 micrometers in diameter. PM_{10} refers to all particles less than or equal to 10 micrometers in diameter. A particle 10 micrometers in diameter is about one-seventh the diameter of a human hair. PM can be emitted directly or form secondarily in the atmosphere. "Primary" particles, such as dust from roads or elemental carbon (soot) from wood combustion, are emitted directly into the atmosphere. "Secondary" particles are formed in the atmosphere from primary gaseous emissions. Examples include sulfate, formed from SO_2 emissions from power plants and industrial facilities; and nitrates, formed from NO_x emissions from power plants, automobiles and other types of combustion sources. The chemical composition of particles depends on location, time of year, and weather. Generally, fine PM is composed mostly of secondary particles, and coarse PM is composed largely of primary particles. . . .

"Between 1991 and 2000, average PM_{10} concentrations decreased 19 percent."

Between 1991 and 2000, average PM_{10} concentrations decreased 19 percent, while direct PM_{10} emissions decreased 6 percent.

Direct $PM_{2.5}$ emissions from man-made sources decreased 5 percent nationally between 1991 and 2000. . . .

Sites in the East typically have higher annual average $PM_{2.5}$ concentrations. Most of the regional difference is attributable to higher sulfate concentrations in the eastern United States. Sulfate concentrations in the eastern sites are 4 to 5 times greater than those in the western sites. Sulfate concentrations in the East largely result from sulfur dioxide emissions from coal-fired power plants. EPA's Acid Rain Program . . . sets restrictions on these power plants. Within the East, rural $PM_{2.5}$ levels are higher in the Southeastern and Mid-Atlantic states. In the West, rural $PM_{2.5}$ levels are generally less than one-half of Eastern levels. . . .

Carbon Monoxide (CO)

Carbon monoxide (CO) is a colorless and odorless gas, formed when carbon in fuel is not burned completely. It is a component of motor vehicle exhaust, which contributes about 60 percent of all CO emissions nationwide. Non-road vehicles account for the remaining CO emissions from transportation sources. High concentrations of CO generally occur in areas with heavy traffic congestion. In cities, as much as 95 percent of all CO emissions may come from automobile exhaust. Other sources of CO emissions include industrial processes, non-transportation fuel combustion, and natural sources such as wildfires. Peak CO concentrations typically occur during the colder months of the year when CO automotive emissions are greater and nighttime inversion conditions (where air pollutants are trapped near the ground beneath a layer of warm air) are more frequent. . . .

Nationally, the 2000 ambient average CO concentration is 61 percent lower than that for 1981 and is the lowest level recorded during the past 20 years. CO emissions levels decreased 18 percent over the same period. Between 1991 and 2000, ambient CO concentrations decreased 41 percent, and the estimated number of exceedances of the national standard decreased 95 percent while CO emissions fell 5 percent. This improvement occurred despite a 24 percent increase in vehicle miles traveled in the United States during this 10-year period.

Lead (Pb)

In the past, automotive sources were the major contributor of lead emissions to the atmosphere. As a result of EPA's regulatory efforts to reduce the content of lead in gasoline, air emissions of lead from the transportation sector have declined over the past decade. Today, industrial processes, primarily metals processing, are the major source of lead emissions to the atmosphere. The highest air concentrations of lead are found in the vicinity of smelters, and battery manufacturers. . . .

Because of the phase-out of leaded gasoline, lead emissions and concentrations decreased sharply during the 1980s and early 1990s. The 2000 average air quality concentration for lead is 93 percent lower than in 1981. Emissions of lead decreased 94 percent over that same 20-year period. Today, the only violations of the lead national air quality standard occur near large industrial sources such as lead smelters. . . .

Toxic Air Pollutants

Toxic air pollutants, or air toxics, are those pollutants that cause or may cause cancer or other serious health effects, such as reproductive effects or birth defects. Air toxics may also cause adverse environmental and ecological effects. EPA is required to reduce air emissions of 188 air toxics listed in the Clean Air Act. Examples of toxic air pollutants include benzene, found in gasoline; perchloroethylene, emitted from some dry cleaning facilities; and methylene chloride, used as a solvent by a number of industries. Most air toxics originate from man-made sources, including mobile sources (e.g., cars, trucks, construction equipment) and stationary sources (e.g., factories, refineries, power plants), as well as indoor sources (e.g., some building materials and cleaning solvents). Some air toxics are also released from natural sources such as volcanic eruptions and forest fires. . . .

> ***"Nationally, the 2000 ambient average CO concentration is 61 percent lower than that for 1981 and is the lowest level recorded during the past 20 years."***

EPA and states do not maintain a nationwide monitoring network for air toxics as they do for many of the other pollutants discussed in this report. Although such a network is under development, EPA has compiled a National Toxics Inventory (NTI) to estimate and track national emissions trends for the 188 toxic air pollutants regulated under the Clean Air Act. In the NTI, EPA divides emissions into four types of sectors: 1) major (large industrial) sources; 2) area and other sources, which include smaller industrial sources, like small dry cleaners and gasoline stations, as well as natural sources, like wildfires; 3) onroad mobile sources, including highway vehicles; and 4) nonroad mobile sources, like aircraft, locomotives, and construction equipment.

Based on 1996 estimates, the most recent year of available data, the emissions of toxic air pollutants are relatively equally divided between the four types of sources. However, this distribution varies from city to city.

While EPA, states and tribes collect monitoring data for a number of toxic air pollutants, both the chemicals monitored and the geographic coverage of the monitors vary from state to state. Together with the emissions data from the NTI, the available monitoring data help air pollution control agencies track trends in toxic air pollutants in various locations around the country. EPA is working with states, tribes and local air monitoring agencies to build upon these monitoring sites to create a national monitoring network for a number of toxic air pollutants.

"Air toxics emissions have dropped approximately 23 percent between baseline (1990–1993) and 1996."

Based on the data in the NTI, estimates of nationwide air toxics emissions have dropped approximately 23 percent between baseline (1990–1993) and 1996. Although changes in how EPA compiled the national inventory over time may account for some differences, EPA and state regulations, as well as voluntary reductions by industry, have clearly achieved large reductions in overall air toxic emissions.

Trends for individual air toxics vary from pollutant to pollutant. For example, data taken from California's monitoring network for 39 urban sites show an average reduction of 60 percent in measured levels of perchloroethylene for the period 1990–1999. Perchloroethylene is a chemical widely used in the dry cleaning industry. Based on the NTI, EPA estimates that nationwide perchloroethylene emissions dropped 67 percent from 1990–1996. These reductions reflect state and federal efforts to regulate emissions of this pollutant, and industry efforts to move to other processes using less toxic chemicals. . . .

Progress Must Continue

The Clean Air Act has resulted in many improvements in the quality of the air in the United States. Scientific and international developments continue to have an effect on the air pollution programs that are implemented by the U.S. Environmental Protection Agency and state, local, and tribal agencies. New data help identify sources of pollutants and the properties of these pollutants. Although much progress has been made to clean up our air, work must continue to ensure steady improvements in air quality, especially because our lifestyles create more pollution sources. Many of the strategies for air quality improvement will continue to be developed through coordinated efforts with EPA, state, local and tribal governments, as well as industry and other environmental organizations.

Factory Farming Is Not Polluting the Water Supply

by Dave Juday

About the author: *Dave Juday is an adjunct fellow at the Hudson Institute's Center for Global Food Issues. The Hudson Institute is a conservative think tank.*

Hog production has been targeted by anti-agriculture environmental groups as a profound threat to the environment. The Environmental Defense Fund (EDF), an advocacy group best-known for its opposition to pesticides, has even launched a "North Carolina Poop Counter" on the Internet to keep a running tally of the pounds of hog manure discharged in the state, and its phosphorous content. North Carolina is the nation's second largest hog-producing state.

A survey conducted in March 1999 by [political research firm] Lake Snell Perry & Associates found that 80 percent of voters surveyed "favor creating tougher, uniform standards to limit the air and water pollution from factory farms." The poll was commissioned by ten anti-farm organizations, including Defenders of Wildlife, the Sierra Club, the Conservation Council of North Carolina, and Ralph Nader's U.S. Public Interest Research Group.

A Porcine Threat to Clean Water?

According to EDF, the "concentration of hogs in eastern North Carolina increases the threats to human health and the environment." The group claims that "the disposal of huge amounts of hog waste is dramatically increasing the nutrient pollution" in the region's surface waters.

According to Carol Browner, [former] administrator of the U.S. Environmental Protection Agency (EPA), livestock waste is a "major source of water pollution." According to the agency's March 1998 report, "Strategy for Addressing Environmental and Public Health Impacts from Animal Feeding Operations," agriculture is the single leading cause of impaired rivers, streams, and lakes, contributing up to 60 percent of the pollution in surveyed rivers and streams.

Farm groups and independent scientific sources dispute much of the case against livestock waste. Observers note that, while hog production in the Black River watershed has increased 500 percent over the past 20 years, the Black River continues to carry an "excellent" rating from the North Carolina Department of Environmental Resources (DNER). Responding to the aforementioned EPA report, Richard Halpern, a former policy planning official from Rockingham County, Virginia, notes "The operative word is surveyed."

Rockingham is the nation's second-largest poultry producing county; Halpern specialized in water quality issues affecting the poultry industry. He is now an adjunct fellow at the Hudson Institute's Center for Global Food Issues, where he's followed closely the EPA's efforts at reforming water quality standards that apply to livestock operations.

Halpern notes that only 17 percent of the nation's river miles have been surveyed, and of that 17 percent, just 37 percent—or 6.3 percent of the nation's total river miles—are known to be impaired. Agriculture is estimated to be responsible for 60 percent of that impairment, with animal feeding operations of all kinds—including dairy, beef feedlots, poultry, and swine—estimated by EPA to "adversely impact 16 percent of those waters."

"In other words," Halpern calculates, "livestock affects 16 percent of 60 percent of 37 percent of 17 percent of the nation's rivers and streams. In the end, that's less than 1 percent total."

The EPA report suffers from a bad case of shoddy data, says Leonard Gianessi, a water quality scientist and the developer of the water quality modeling method used to measure feedlot and confinement livestock waste. He says the model's assumption were "very crude" a decade-and-a-half ago. "The really ridiculous part is that these numbers get encoded in all of these resource assessments as though the model is really accurate. Fifteen years later, and the (government) is still using these numbers. They weren't that good to begin with," according to Gianessi.

In Congress, a bipartisan group of farm state representatives wrote [former] EPA Administrator Browner and [former] U.S. Department of Agriculture Secretary Dan Glickman, expressing their reservations about many of the assumptions in the EPA's livestock waste report. Echoing Gianessi's concerns, the congressmen also challenge a number of the agency's recommendations.

"Currently much of the information and statistics on water quality in the U.S. are incomplete and/or so dated that they can no longer be considered accurate," they write. "States are presently working to address (livestock) issues, and the EPA and USDA should give these efforts a chance to work before additional regulations are promulgated."

"Farm groups and independent scientific sources dispute much of the case against livestock waste."

The livestock waste controversy is no less heated at the state level. In Maryland, for example, legislation controlling the use of poultry manure as fertilizer cleared the legislature and was signed into law based on press reports linking the toxic microbe *Pfiesteria Piscicida* to nutrient pollution from poultry production. Yet the governor's "Blue Ribbon Citizens' Pfiesteria Action Commission," after studying the existing scientific research, reported that there is no "demonstrable cause and effect linkage between farm runoff and Pfiesteria."

In November 1998, South Dakota voters approved a constitutional amendment—widely known to be aimed at corporate hog farms—that not only prohibits corporations from owning farmland in South Dakota, but also ends the practice of companies contracting with farmers to raise crops or livestock on their behalf. Seven states have laws aimed at restricting corporate involvement in farming, but only Nebraska has joined South Dakota in writing a constitutional ban.

In Iowa—the country's number one pork producer—livestock operators are presumed by law not to be a nuisance. In 1998, the state legislature passed a bill that preempts county officials from regulating animal feeding operations.

Hope in Technology

A recent scientific breakthrough in Canada may reshape the debate over livestock waste and water quality.

Reuters reports that scientists at the University of Guelph in Ontario have developed a strain of pigs that has the potential to produce manure 20 to 50 percent lower in phosphorus content. Phosphorus is one of the nutrients present in livestock manure which, along with nitrogen, has been blamed by environmental groups and some policy makers as a threat to water quality in the U.S. and Canada.

"'Livestock [waste] affects . . . less than 1 percent [of the nation's rivers and streams].'"

According to the Reuters story, three Yorkshire pigs—individually named after famous hockey stars, including the recently retired Wayne Gretzky—have been born to surrogate sows. Collectively, the new porcine strain is called the Enviropig, and each of their cells' nuclei includes DNA from mice and bacteria that regulate the phosphorus in their waste.

The genetic engineering research at the University of Guelph, funded by the Ontario Pork Council, is considered to be the first using biotechnology to genetically engineer animals to solve environmental problems. Coupled with other efforts underway to reduce the phosphorus content in feed grains, these technology breakthroughs could indeed cast the livestock waste debate in a new light.

If nutrients such as phosphorus are reduced even before being excreted in livestock manure, the problem with pigs—to the extent a problem exists at all—will largely disappear. Anti-agriculture groups such as EDF will have to search for a new target.

New Coal-Burning Technology Is Less Polluting

by Robert S. Kripowicz

About the author: *Robert S. Kripowicz is the acting assistant secretary for fossil energy for the U.S. Department of Energy.*

Coal is an indispensable part of our nation's energy mix. Because of its abundance and low cost, coal now accounts for more than half of the electricity generated in this country.

Coal is our nation's most abundant domestic energy resource. One quarter of all the world's known coal supplies are found within the United States. In terms of energy value (Btus), coal constitutes approximately 95 percent of U.S. fossil energy reserves. Our nation's recoverable coal has the energy equivalent of about one trillion barrels of crude oil—comparable in energy content to all the world's known oil reserves. At present consumption rates, U.S. coal reserves are expected to last at least 275 years.

Coal has also been an energy bargain for the U.S. Historically it has been the least expensive fossil fuel available to the country, and in contrast to other primary fuels, its costs are likely to continue to decline as mine productivity continues to increase. Between 1988 and 1997, minemouth coal prices (in real 1992 dollars) declined by $9.40 per ton, or 37 percent; between 1998 and 2020, prices could decline by another $5.00 per ton (1998 dollars), or about 1.5 percent a year. The low cost of coal is a major reason why the United States enjoys some of the lowest electricity rates of any free market economy.

Coal Consumption and Demand for Electricity

America's coal industry—81,000 miners working in 25 states—produces approximately 1 billion tons of coal per year. Just under 950 million tons goes to U.S. power plants (the rest is used for industrial purposes, such as steelmaking,

Excerpted from Robert S. Kripowicz's testimony before the Subcommittee on Energy and Power, Committee on Commerce, U.S. House of Representatives, June 8, 2000.

or is exported). According to the Department of Energy's (DOE) Energy Information Administration (EIA), domestic coal demand could increase by 20 percent by 2020, growing to 1,316 million tons, primarily because of increasing coal use for electricity generation. Although coal's overall contribution to the nation's electric power supply is projected to decline somewhat—from 52 percent in 1998 to 49 percent in 2020—the substantial growth in U.S. power consumption means that the U.S. will mine and use more coal in the foreseeable future.

A key element in EIA's projection is that very little new capacity is planned during that time period, about 7% of existing capacity (or around 21 gigawatts). Most of the increased generation from coal-fired units will come from existing plants increasing their hours of operation. The primary barrier to construction of new coal-fired power plants will be intense competition from natural gas combined cycle power plants. These natural gas–fired plants have much lower capital costs than coal plants and are very low pollutant emitters. Electricity restructuring is another important development in the industry. Using authorities provided by Congress in the Energy Policy Act of 1992 and other statutes, the Federal Energy Regulatory Commission has taken action to make wholesale electricity markets more competitive. To date, 25 states have taken action to introduce competition into retail electricity markets and many others are considering this option. . . .

Coal and the Environment

Largely because of improving pollution control technology, the nation has been able to use more coal while improving the quality of its air. Coal use has more than doubled since 1970 while emissions of sulfur and nitrogen pollutants have declined by 70 percent and 45 percent respectively.

EIA's coal projections reflect existing environmental regulations only. Whether expectations for future growth in coal demand actually materialize will depend largely on the nation's coal users' ability to comply with increasingly stringent environmental regulations. Increased compliance costs can lead to early retirement of a unit, or to less use of the coal-fired generating unit as it becomes more costly to operate. The most critical regulations and policy initiatives are air-pollution related and include:

"At present consumption rates, U.S. coal reserves are expected to last at least 275 years."

Rules to address the Regional Transport of Ozone (the ozone "SIP Call" and related rules promulgated by EPA). The SIP Call rule required 22 Eastern states and the District of Columbia to reduce nitrogen oxide (NO_x) emissions by specified amounts by May 2003. Although the rules are being revised to comply with judicial direction, the primary mechanism to achieve the required reductions is expected to be additional NO_x reduction requirements at coal-fired power plants.

Revised National Ambient Air Quality Standards for Particulate Matter and for Ozone. These revised standards were promulgated in 1997, with anticipated annual compliance costs for full attainment of $37 billion per year and $10 billion per year, respectively. The Supreme Court will be reviewing the Environmental Protection Agency (EPA) rules. Both are significant for power plants because they will lead to additional reductions in emissions of NO_x and sulfur dioxide (SO_2) which are precursors to fine airborne particles.

Mercury regulations. Under a court sanctioned agreement, EPA is scheduled to decide by December 15, 2000, whether or not it is necessary to control mercury from coal-fired power plants. If EPA deems it necessary, the agency must promulgate regulations by December 2003.

Enforcement initiative. On November 3, 1999, EPA filed lawsuits against seven utility companies, and issued an administrative order against an eighth, charging violation of new source review requirements. The civil actions, now in the discovery stage, all seek retrofit of state-of-the-art control technology. A total of 33 gigawatts of capacity is involved in EPA's initiative—over 10% of total U.S. coal-fired capacity. The basic allegation is that activities at these plants were modifications requiring new source permits. In the only settlement to date, the Tampa Electric Company (TECO) agreed to 85% reductions in NO_x and SO_2 by 2010, retirement of significant coal capacity, and payment of a $3.5 million civil penalty.

"The nation has been able to use more coal while improving the quality of its air."

The 305 gigawatts of existing coal-fired power plants can be categorized into three groups: (1) very large and relatively new plants, (2) very small and relatively old plants, and (3) those in between. The first category will probably be able to continue to operate economically, even with the new regulations. Many of the smaller plants in the middle category will not, and in fact several utilities have recently announced plans to replace some older coal units with new natural gas–fired units.

The pivotal group is the third group—moderate size coal plants with significant remaining operational lifetimes. It is this group which will benefit most from development and deployment of advanced emission control technologies. The greater the success of DOE and its private sector partners in developing more effective, and lower cost mitigation technologies, the more of these plants which will continue to operate, and the lower the overall cost of electric power will be to the consumer.

A major caveat is that none of the projections assumes the implementation of new regulation to address climate change concerns. DOE is also pursuing technologies to reduce greenhouse gas emissions from coal (and natural gas) power plants—both by increasing efficiency of the power generating process and by capturing and sequestering carbon gases. Although these technologies are longer term and unlikely to be available prior to 2015, they could allow for the use of coal as a fuel for new generating plants while substantially reducing or even eliminating emissions of greenhouse gases to the atmosphere.

Measures to reduce greenhouse gas emissions before 2015 could lead to significant reductions in domestic coal use. Impacts on domestic coal use would likely be directly related to the amount of reduction in greenhouse gas emissions that takes place within U.S. borders. For a given level of greenhouse gas emissions commitment, provisions that allow the U.S. to meet the commitment by (1) relying on purchased emissions reductions from sources in other countries, (2) sequestration of carbon dioxide through forestry activities, and (3) additional reductions of non–carbon dioxide greenhouse gases would reduce the impact of any such obligation on the level of domestic coal use.

> ***"In the mid-1980s, the United States began an unprecedented . . . investment in a new generation of cleaner coal technologies."***

Clean Coal Technology—the Investment Is Paying Off

With coal expected to remain one of the nation's lowest cost energy sources, its future will be determined largely by the availability of affordable technology that can reduce the impact of its use on the environment.

In the mid-1980s, the United States began an unprecedented joint public-private investment in a new generation of cleaner coal technologies. The Clean Coal Technology Program led to 40 projects in 18 states, over half successfully completed. More than $5.6 billion has been committed to this program, with private industry and states investing two dollars for every one from the federal government. Today, because of the Clean Coal Technology Program and the research efforts that undergird it, pollution control costs are significantly lower. In the mid-1980s, the only options to reduce smog-causing nitrogen oxide (NO_x) pollutants from coal-fired power plants cost $3,000 per ton of NO_x. Today, technologies such as low-NO_x burners demonstrated in the Clean Coal Technology Program have reduced NO_x control costs to less than $200 per ton. Nearly 75 percent of the nation's coal-fired generating capacity now uses low-NO_x burners. The cost of selective catalytic reduction, which removes NO_x from coal flue gases, has been cut in half because of technology advances.

Similarly, in the 1970s, scrubbers—the flue gas treatment devices that remove sulfur pollutants from the exhausts of coal-fired boilers—were expensive, unreliable, and posed waste handling problems. The Federal Government's [research and development] program (both at DOE and EPA) and DOE's Clean Coal Technology Program helped improve scrubber technologies. Today, flue gas scrubbers are one-fourth as expensive as the vintage-1970s units and operate much more reliably. The reduced costs, alone, have saved American ratepayers more than $40 billion since 1975. Today, advanced scrubbers produce a waste product that can be recycled into wallboard or easily disposed of in a safe, powder form, rather than the sludge of older systems.

Coal Combustion Is Cleaner

In the 1970s and 80s, DOE's R&D program helped develop the fluidized bed coal combustor—an advanced coal-burning technology that removed sulfur pollutants and limited the formation of NO_x pollutants inside the boiler, eliminating the need for scrubbers or other post-combustion controls. The new technology found widespread acceptance in the industrial boiler market. The Clean Coal Technology Program helped move this clean-burning technology into the larger-size, utility market. Using this technology, coal-fired power plants can reduce sulfur emissions by more than 95 percent and NO_x emissions by more than 90 percent, even when burning high-sulfur coal.

Utilities have a new option for coal-based power. The Clean Coal Technology Program also pioneered a fundamentally new way to use coal to generate electricity. Rather than burning it in a boiler, gasification-combined cycle technology first converts coal into a combustible gas, cleans the gas of virtually all of its pollutants, then burns the gas in a turbine, much like natural gas. More than 99 percent of sulfur, nitrogen, and particulate pollutants can be removed in the process.

> ***"Using [clean-burning] technology, coal-fired power plants can reduce sulfur emissions by more than 95 percent."***

Moreover, heat from the turbine can be used in a conventional steam cycle to generate a second source of electricity, increasing overall power plant efficiencies.

Because of the Clean Coal Technology Program, the nation now has three full-scale, pioneering coal gasification combined cycle power plants located in Florida, Indiana, and Nevada. These are among the cleanest fossil fuel power generating facilities in the world.

Steel mills have an environmentally attractive alternative to coke ovens. Much of the nation's coal not used by power plants is shipped to steel mills for use in making the coke needed for the steel making process. Coke production, however, is a significant source of air pollutants, including air toxics. The Clean Coal Technology Program demonstrated a way to use coal directly in the blast furnace, displacing coke virtually on a pound-for-pound basis. Direct coal injection offers the steel industry a clearly superior economical and environmental alternative to traditional coke-making.

The Future

When the Department of Energy (DOE) issued the Comprehensive National Energy Strategy in April 1998, the first of its five overarching goals was to improve the efficiency of the energy system—making more productive use of energy resources to enhance overall economic performance while protecting the environment . . .

One of the major strategies to achieve this goal is to demonstrate cost-effective power systems that can achieve electrical generating efficiencies greater than 60 percent. Today's coal-fired power plants convert only about a third (between 33–35 percent) of the energy value of coal into electricity. The rest is typically discarded as waste heat. The Clean Coal Technology Program has demonstrated new technologies that can boost efficiencies to nearly 45 percent. Advances now in the DOE research and development program—for example, more energy-efficient gas separation technologies, improved turbines, and coal-capable fuel cells—could push coal power plant efficiencies into the 60 percent range.

"A 60 percent efficient coal power plant can cut carbon dioxide emissions by more than 40 percent."

More Efficient Coal-Fired Power Plants

Cleaner operation is one, since a coal plant that uses less fuel to generate the same amount of power will emit fewer emissions. Reduced greenhouse gas emissions is another benefit; a 60 percent efficient coal power plant can cut carbon dioxide emissions by more than 40 percent. A third is cost to consumers. Improving the efficiency of a power plant can lower costs of the electricity generated, perhaps by up to 20 percent.

It may be possible in the future to eliminate virtually all of the environmental concerns at a coal-based power plant.

DOE is developing a concept for a new fleet of energy facilities that would incorporate breakthrough technologies in advanced power generation and pollution controls. With a target date of 2015, this new energy concept, called Vision 21, would incorporate technologies that would reduce SO_2 (sulfur dioxide) and NO_x emissions to near zero, and cut in half the amount of carbon dioxide emitted from the plant.

Future coal-fired energy facilities may look radically different from today's power plants. One possible configuration for DOE's Vision 21 concept envisions a facility that could process both coal and municipal wastes so cleanly that it could be sited near major urban centers. . . .

Moreover, the Vision 21 concept could incorporate various coproduction options—producing not only electricity but other high-value products such as hydrogen, clean transportation fuels, chemicals and other commercial commodities. By developing a multi-product energy facility—rather than just a single-product electrical generating plant—it may be possible to boost overall coal use efficiencies to more than 80 percent.

Improving the efficiency of tomorrow's coal-fueled energy facilities can be a beneficial companion to improving end-use energy conservation efforts. For example, by raising the efficiency of U.S. coal-fired power plants to 50 percent, the nation could achieve fuel savings equivalent to weatherizing 400 million homes—more than 5 times the number of homes in the United States.

Carbon Sequestration and Greenhouse Gas Reductions

Even with improved efficiencies, a future coal-fired power plant still may not be able to achieve the substantial greenhouse gas reductions that may be necessary to counter concerns about global climate change. Therefore, one of the keys to coal's long-term future (and to the future of other fossil fuels) may be the emerging technology of carbon sequestration.

Only a few years ago, concepts for capturing greenhouse gases at their point of emission, or even from the ambient air, and either storing them for centuries or recycling them into useful products were considered laboratory curiosities. Today, the opinion is much different.

DOE has set a goal of developing technologies that can capture and sequester carbon dioxide at costs as low as $10 per ton of carbon. This is equivalent to adding only 2/10ths of a cent per kilowatt-hour to electricity rates that today range from 4 to 12 cents per kilowatt hour.

Carbon sequestration—if the technology can be successfully developed—could be the only option that doesn't require large-scale turnover of the world's energy infrastructure. Along with low-carbon and carbon-free energy supply technologies, such as natural gas and renewable energy systems, and more energy-efficient end-uses, carbon sequestration could become an important 3rd option in reducing the buildup of greenhouse gases.

The United States needs a variety of energy sources to continue the unprecedented economic expansion that has made us the envy of the world. At the same time, Americans have consistently ranked environmental quality as one of their highest priorities for both current and future generations.

While the U.S. will continue to expand the role of renewable and other alternative energy resources in its energy portfolio, coal will continue to provide a large share of the overall energy—and the dominant share of electricity—that can keep our economy growing. New technologies can make it possible to use all of our domestic energy resources—including our largest resource, coal—in ways that are compatible with our goals to protect the environment. Over the past 20-year history of the Department of Energy, we have made substantial progress in improving the environmental acceptability of coal use while, at the same time, keeping the costs of coal-derived energy low. Through the continued public and private investment into advanced, more efficient, and cleaner coal technologies, coal can remain a beneficial contributor to America's energy future.

Air Pollution from Automobiles Has Been Reduced

by Joseph L. Bast and Jay Lehr

About the authors: *Joseph L. Bast is president of the Heartland Institute, a conservative research and education organization. Jay Lehr is science director for the Heartland Institute and senior scientist of Environmental Education Enterprises, a producer of continuing education courses for environmental professionals.*

Air pollution is generally counted as a cost, or "negative externality," of popular ownership of cars and trucks. Often missing from such calculations, however, are the *beneficial effects on human health and the environment* delivered by cars and trucks when they replaced prior modes of travel and cartage that polluted more. "The modern American automobile," writes [public policy journalist] Gregg Easterbrook, "is the cleanest system of transportation ever devised."

Environmental Benefits of the Automobile

Before the advent of motorized travel, most Americans relied on horses or sat in carriages pulled by horses for personal transportation. Freight also moved by horse-drawn wagons or, for longer distances, smoke-belching trains. Fred L. Smith, a former senior policy analyst for the Environmental Protection Agency (EPA) and now president of the Competitive Enterprise Institute, gives this vivid and unpleasant description of the impact horses had on America's cities 100 years ago:

> A horse produces approximately 45 pounds of manure each day. In high-density urban environments, massive tonnages accumulated, requiring constant collection and disposal. Flies, dried dung dust, and the smell of urine filled the air, spreading disease and irritating the lungs. On rainy days, one walked through puddles of liquid wastes. Occupational diseases in horse-related industries were common.

Smith goes on to report that New York City in the 1890s had to dispose of

Excerpted from "The Increasing Sustainability of Cars, Trucks, and the Internal Combustion Engine," by Joseph L. Bast and Jay Lehr, *Heartland Policy Study*, June 22, 2000.

15,000 dead horses every year, a huge public health and environmental problem. Often, these rotting corpses were hauled in open-air wagons to the edge of town, where they were dumped into huge kettles and heated over coal fires (without emission controls) until they were "reduced" into grease, later to be sold to the manufacturers of candles and lubricants. The remains of the dead animals that couldn't be sold were dumped, untreated, into the nearest river or lake.

The horses and mules that moved freight and provided automobility in the nineteenth century needed to be fed, and here we see another major negative environmental effect. Approximately 93 million acres of land in the U.S. were devoted to horse grazing in 1915, an area nearly one-third larger than all the cities in the U.S. today. By 1961, thanks to popular ownership of cars and trucks, the area needed to feed horses had shrunk so low that the government stopped keeping records on it. Much of that land has since returned to forest.

What would have been the effect on human health and the environment if cars and trucks hadn't emerged as a replacement for horses and mules? We would not have turned to trains for most of our transportation needs, first because they lack even the speed and flexibility of horses and mules for short trips, and second because the nation could not possibly have afforded to build and operate train lines reaching all the areas where a surging population wanted to live, work, and visit. Population growth and rising wealth, which historically have increased preferences for larger homes, lower density in residential areas, and more travel, would have greatly boosted demand for horses and mules.

One can only wonder at the enormous quantities of horse manure and urine that would have been deposited every day on city streets and the terrible health consequences of the same, the tremendous challenge of disposing of millions of dead horses each year, and the massive deforestation of the continent that would have been required to feed a vastly enlarged herd of horses and mules, if not for the timely arrival of cars and trucks around the turn of the century.

Cars and trucks, in other words, arrived just in time to rescue the nation's growing cities from an ecological catastrophe. Air quality and threats to public health in major cities (and to a lesser extent, smaller ones, too) would unquestionably be far worse today if we still relied on horses and mules for personal transportation and cartage. Deforestation would have taken place on an unprecedented scale to feed those horses and mules. . . . Cars and trucks, while not entirely free of environmental complications themselves, made major positive contributions to environmental protection. . . .

"'The modern American automobile . . . is the cleanest system of transportation ever devised.'"

There is little evidence that exposure to current levels of air pollution is a risk factor for cancer or other health problems. What small health effects might have occurred in the past are likely to have disappeared in recent years since air quality in nearly all major American cities has improved. Ambient air concentrations of all six of the air pollutants tracked by the Environmental Protection Agency (EPA) fell between 1978 and 1997: Sulfur dioxide concentrations fell by 55 percent; carbon monoxide, by 60 percent; ozone, by 29.5 percent; nitrogen dioxide, by 25 percent; and lead, by 96.5 percent. Particulate matter (PM) has fallen 25 percent since 1988, the year after EPA changed its methodology for measuring total national emissions of PM. According to EPA, "PM-10 emissions from on-road vehicles and non-road sources have declined approximately 73 percent during the 1940 to 1994 period."

"Cars and trucks . . . arrived just in time to rescue the nation's growing cities from an ecological catastrophe."

Another way to view improvements in air quality is to observe trends in the number of days per year that ambient levels of ozone exceeded federal air quality standards. Since meteorological conditions influence the creation of ozone and vary from year to year, a good comparison is between the six-year periods 1987–1992 and 1994–1999. Such comparisons show the number of "bad air" days fell 76 percent in Boston, 78 percent in Chicago, 54 percent in Los Angeles, 69 percent in Milwaukee, 82 percent in Newark, and 88 percent in San Diego.

Strengthening Vehicle Emission Standards

Improvements in vehicle emission controls helped drive this improvement in air quality. A new car produced in 1997 or later produces 98 percent fewer hydrocarbons, 96 percent less carbon monoxide, and 89 percent fewer nitrogen oxide emissions per mile than a new car produced before 1975.

Emission standards for heavy-duty trucks, typically diesel-powered, were greatly strengthened in 1995, requiring reductions from precontrol baseline emissions of 94 percent for hydrocarbons, 89 percent for carbon monoxide, 64 percent for nitrogen oxides, and 83 percent for diesel particulates. . . . EPA [has] announced new regulations on diesel trucks and diesel fuel aimed at a further 90 percent reduction in emissions.

The progress made so far in making city air cleaner and safer will continue well into the twenty-first century. Gradual replacement of older cars and trucks with vehicles equipped with new pollution prevention devices will reduce emissions by retiring the worst polluters in the fleet. New emission standards and clean fuels mandates . . . are being phased in, along with controls on electricity generating plants and other stationary sources of air pollution.

The migration of jobs from central cities to suburban locations works in favor of cleaner air in two ways: first by lowering the concentration of firms that release manufacturing-related emissions, and second by replacing suburb-to-city commuting with shorter suburb-to-suburb commutes. The ongoing transition of the U.S. economy from a preponderance of manufacturing jobs to service jobs means fewer emissions at work as well as more opportunities for telecommuting and less work-related travel.

A study conducted for the Foundation for Clean Air Progress projected to the year 2015 emissions of the six air pollutants regulated by EPA. The study used Department of Energy and EPA projections of future energy consumption and emissions and assumed that current air quality laws would continue to be phased in but no new air quality regulations would be enacted. The forecast: Total emissions will fall by 22 percent between 1997 and 2015. Emissions in 2015 are projected to be over 32 million tons less than in 1997, and 109 million tons less than in 1970.

While forecasting the future by looking at past trends is fraught with peril, . . . structural changes are taking place in our economy, our cities, and our cars and trucks that will work to secure past gains in air quality and carry them forward into the twenty-first century. Gregg Easterbrook, writing in 1995, saw this convergence of environmentally friendly trends and declared that "the Age of Pollution is nearly over. Almost every pollution issue will be solved within the lifetimes of readers of this book."

Automobiles and the Global Warming Debate

Scientists have discovered that concentrations of trace greenhouse gases in the atmosphere, particularly carbon dioxide (CO_2), are rising. Theoretically, these gases could trap more heat in the atmosphere, leading to a gradual warming of the Earth's atmosphere. Cars and trucks are part of the global climate change debate because CO_2 is released whenever fossil fuels are burned. Cars and trucks account for about 25 percent of U.S. fossil CO_2 emissions.

In 1997, representatives of the United States and other nations met in Kyoto, Japan, to negotiate an amendment to an earlier treaty, called the Rio Treaty, to address the possible threat of global climate change. That amendment, called the Kyoto Protocol, would require the U.S. to reduce its greenhouse gas emissions—primarily carbon dioxide (CO_2), methane (CH_4), and nitrous oxide (N_2O)—to 7 percent below 1990 levels by the year 2012. . . . [Editor's note: President George W. Bush has withdrawn U.S. support for the Kyoto Protocol as the treaty is currently written.]

> ***"There is little evidence that exposure to current levels of air pollution is a risk factor for cancer or other health problems."***

There is considerable scientific dispute over whether the threat of anthropogenic global climate change is real. Over 17,000 scientists have signed a petition saying, in part, "there is no convincing scientific evidence that human release of carbon dioxide, methane, or other greenhouse gases is causing or will, in the foreseeable future, cause catastrophic heating of the Earth's atmosphere and disruption of the Earth's climate."

Over 100 climate scientists have signed the Leipzig Declaration, which states in part, "there does not exist today a general scientific consensus about the importance of greenhouse warming from rising levels of carbon dioxide. On the contrary, most scientists now accept the fact that actual observations from Earth satellites show no climate warming whatsoever." A 1997 survey of state climatologists found that 89 percent agreed that "current science is unable to isolate and measure variations in global temperatures caused only by man-made factors."

Figure 1. Selective Use of Data in Intergovernmental Panel on Climate Change (IPCC) Report

Solid circles indicate years studied by Santer et al. for their chapter in the IPCC report. Open circles complete the data series and show no warming trend.

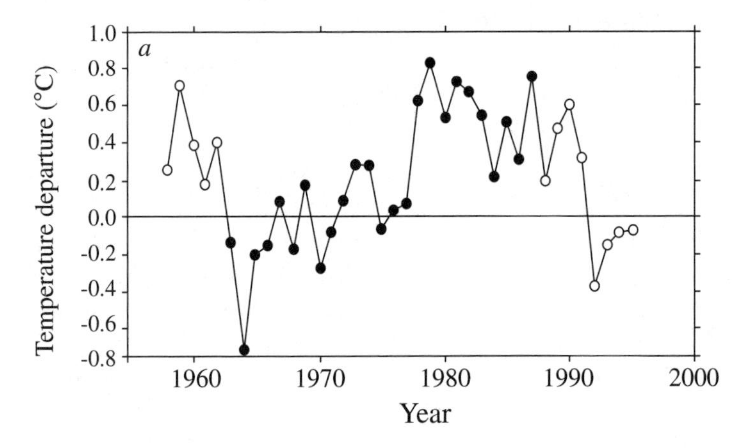

P.J. Michaels and P.C. Knappenberger, "Human Effect on Global Climate?" *Nature,* vol. 384, December 12, 1996, p. 522.

Dr. Benjamin Santer, the lead author of the science chapter of the widely cited report by the Intergovernmental Panel on Climate Change (IPCC), *Climate Change 1995*, coauthored an article for a peer-reviewed scientific journal around the same time as the IPCC report was written. In that essay, Santer and his colleagues say it is not possible to get the general circulation models to replicate the past climate record, and until this is resolved, "it will be hard to say, with confidence, that an anthropogenic climate signal has or has not been detected."

It has also been pointed out that the warming trend that formed the basis of Santer's IPCC chapter "is largely an artifact of the time period chosen. . . . Although there is a statistically significant warming in the period from 1963–1987, there is no significant change in the entire (1958–1995) record." Figure 1 makes it apparent that had Santer et al. used the entire time series available to them, their chapter in the IPCC report would not have been cited as proof of a "discernible human influence" on global climate patterns.

More recently Santer has said, "It's unfortunate that many people read the media hype before they read the chapter. . . . I think the caveats are there. We say quite clearly that few scientists would say the attribution issue was a done deal." In a June 2, 1997 debate, IPCC chairman Dr. Bert Bolin said, "the climate issue is not 'settled'; it is both uncertain and incomplete."

"Imperceptible" Car and Truck Emissions

If the science of global climate change were somehow to become more certain and the onset of a genuine problem confirmed, would limiting car and truck emissions make a difference? Carbon dioxide and other greenhouse gases are called "trace gases" because together they constitute just 1 percent of the atmosphere and account for just 3 percent of the natural greenhouse effect. Water vapor accounts for the remaining 97 percent.

CO_2 emissions from cars and trucks in the U.S. amounted to 281 million tons in 1995, about 4 percent of total human greenhouse gas emissions and just 0.18 percent of worldwide emissions from all sources, human as well as natural. Eliminating the emissions of every car and truck in the U.S. (and replacing them with nothing), then, would reduce greenhouse gas emissions worldwide by an almost imperceptible 0.18 percent. Small changes in the amount of water vapor in the atmosphere, in nature's output of CO_2 or its ability to absorb the same, or myriad other poorly measured and understood variables could easily offset the relatively tiny contribution to atmospheric levels of greenhouse gases made by cars and trucks each year.

Fear of anthropogenic global warming should not be a threat to continued reliance on fossil fuels and internal combustion engines, nor should the CO_2 emissions of cars and trucks be considered a "negative externality" that justifies higher taxes or other penalties. Most scientists believe we do not understand global climatic processes well enough to forecast future climate changes. Even those who are most identified with the theory of anthropogenic global warming have admitted that the subject is too complex to make confident predictions.

Chapter 2

Are Corporations Polluting the Environment?

Chapter Preface

On the night of December 2, 1984, the methyl isocyanate (MIC) plant of the Union Carbide Corporation chemical factory in Bhopal, India, severely malfunctioned during routine maintenance. The resulting explosion at the plant dispersed clouds of MIC—a toxic chemical used in the production of insecticides and herbicides—combined with other poisonous chemicals over the city. According to some estimates, six thousand men, women, and children were killed within the first three days of the accident, while deaths to date are estimated at over twenty thousand. The Indian Council of Medical Research has concluded that over half a million people were exposed to the toxic chemicals, many of whom suffered permanent damage to their lungs, eyes, muscles, and reproductive systems. Critics charged that Union Carbide was negligent in maintaining the plant's safety systems, and the company settled a suit with the government of India in 1989, agreeing to pay $470 million on behalf of the victims.

The Bhopal chemical leak, the worst industrial disaster to date, was a public relations nightmare for Union Carbide, as people around the world began to associate the company with the deaths of thousands. The incident prompted people to question whether corporations in general were acting to protect human health and the environment. According to Gary Cohen of the Military Toxics Project, "In the United States, the Bhopal disaster was the impetus for the passage of the Community Right to Know and Emergency Planning Law in 1986, which for the first time gave citizens critical information about a corporation's toxic chemical releases and required companies and local officials to plan on how to address a potential 'Bhopal-like' situation." In the post-Bhopal era, corporations became more accountable to the public for disclosing the potential harm of their operations.

Since the mid-1980s, many large corporations have responded to greater public scrutiny by voluntarily engaging in business practices less damaging to the environment. According to Claudia H. Deutsch, a journalist for the *New York Times*, "[Communities] want [companies] to set aside wilderness areas, clean up rivers that they never had a hand in soiling and be far more squeaky-clean than the government insists. And companies are loath to fight back." Public pressure has created a new climate in which few corporations wish to appear insensitive to environmental concerns. Questions remain, however, over the extent to which corporations are truly committed to placing the environment before profits. The viewpoints in the following chapter debate whether corporations are actually reducing pollution or simply manipulating public sentiment to make it appear that they are.

Corporations Are Reckless Polluters

by Russell Mokhiber

About the author: *Russell Mokhiber is editor of* Corporate Crime Reporter, *an anticorporate legal weekly based in Washington, D.C. He is also the author of* Corporate Crime and Violence.

As we move to the end of the millennium, it is important to remind ourselves that this has been the century of the corporation, where for-profit, largely unaccountable organizations with unlimited life, size and power, took control of the economy and the political economy—largely to the detriment of the individual consumer, worker, neighbor and citizen.

Unspeakable Corporate Damage

Let us again remind ourselves that corporations were created by the citizenry. (Thanks here to Richard Grossman and the Project on Corporations Law and Democracy for resurrecting and teaching us a history we would have collectively forgotten.)

In the beginning, we the citizenry created the corporation to do the public's work—build a canal or a road.

We asked people with money to build the canal or road. If anything went wrong, the liability of these people with money—shareholders, we call them—would be limited to the amount of money they invested and no more. This limited liability corporation is the bedrock of the market economy. The markets would deflate like a punctured balloon if corporations were stripped of limited liability for shareholders.

And what do we, the citizenry, get in return for this generous public grant of limited liability? Originally, we told the corporation what to do. Deliver the goods. And then let humans live our lives.

But corporations gained power, broke through democratic controls, and now roam around the world inflicting unspeakable damage on the earth.

Excerpted from "The Ten Worst Corporations of 1999," by Russell Mokhiber, *Multinational Monitor*, December 1999.

Let us count the ways: price-fixing, chemical explosions, mercury poisoning, oil spills, destruction of public transportation systems. Need concrete examples? These could be five of the most egregious of the century:

Number five: Archer Daniels Midland (ADM) and price fixing. In October 1996, Archer Daniels Midland (ADM), the good people who bring you National Public Radio, pled guilty and paid a $100 million criminal fine—at the time, the largest criminal antitrust fine ever—for its role in conspiracies to fix prices to eliminate competition and allocate sales in the lysine and citric acid markets worldwide.

Federal officials said that as a result of ADM's crime, seed companies, large poultry and swine producers and ultimately farmers paid millions more to buy the lysine additive.

In addition, manufacturers of soft drinks, processed foods, detergents and others, paid millions more to buy the citric acid additive, which ultimately caused consumers to pay more for those products.

Number four: Union Carbide and Bhopal. In 1984, a Union Carbide pesticide factory in Bhopal, India released 90,000 pounds of the chemical methyl isocyanate. The resulting toxic cloud killed several thousand people and injured hundreds of thousands.

Several years of litigation in India resulted in a payment of $470 million by Union Carbide.

In October 1991, the Indian Supreme Court held that the criminal investigation and prosecution of Union Carbide should proceed and stated that failure to accomplish this would constitute "a manifest injustice."

> ***"Corporations gained power, broke through democratic controls, and now roam around the world inflicting unspeakable damage."***

Although Union Carbide was a party to all of these proceedings, it subsequently refused to comply with all efforts to obtain its appearance for the criminal trial by the Bhopal District Court. The efforts of Indian authorities to secure jurisdiction over Union Carbide—including the service of summons on Union Carbide through the U.S. Department of Justice and INTERPOL [the International Criminal Police Organization]—have proved futile.

Mercury, Oil, and Diesel

Number three: Chisso Corporation and Minamata. Minamata, Japan was home to Chisso Corporation, a petrochemical company and maker of plastics. In the 1950s, fish began floating dead in Minamata Bay, cats began committing suicide and children were getting rare forms of brain cancer.

The company had been dumping mercury into the bay, a fact which it at first denied.

By 1975, Chisso had paid $80 million to the 785 verified victims of what became known as Minamata disease. Thousands of other residents claimed they were affected, but were denied compensation.

Number two: Exxon Corporation and Valdez Oil Spill. The Exxon Valdez hit a reef in Prince William Sound Alaska and spilled 11 million gallons of crude oil onto 1,500 miles of Alaskan shoreline, killing birds and fish, and destroying the way of life of thousands of Native Americans.

Most people believe that the Valdez ran aground because the skipper was drunk. Well, he was drunk, but he was also asleep in his bunk, and his third mate was at the wheel. And the third mate was effectively driving blind, as his Raycas radar had been out of order for months.

"In 1984, a Union Carbide pesticide facility in Bhopal, India released 90,000 pounds of [pollutants and] . . . killed several thousand people."

In March 1991, Exxon Corporation and Exxon Shipping pled guilty to federal criminal charges in connection with the March 24, 1989 Valdez oil spill and were assessed a $125 million criminal fine.

The companies pled guilty to misdemeanor violations of federal environmental laws.

Number one: General Motors (GM) and the Destruction of Inner City Rail. Seventy years ago, clean, quiet efficient inner city rail systems dotted the U.S. landscape. The inner city rail systems were destroyed by those very companies that would most benefit from their destruction—oil, tire and automobile companies, led by General Motors.

By 1949, GM had helped destroy 100 electric trolley systems in New York, Philadelphia, Baltimore, St. Louis, Oakland, Salt Lake City, Los Angeles and elsewhere.

In April 1949, a federal grand jury in Chicago indicted and a jury convicted GM, Standard Oil of California and Firestone, among others, of criminally conspiring to replace electric transportation with gas and diesel-powered buses and to monopolize the sale of buses and related products to transportation companies around the country.

GM and the other convicted companies were fined $5,000 each.

And these are not unusual examples. Books have been written documenting the destruction. The question remains—how do we put a stop to it?

And the answer seems clear—reassert public control over what was originally a public institution.

The ideas on how to reassert such control are the subject of debate and conflict, in Seattle and around the world. But as the twentieth century was the century of the corporation, the twenty-first promises to be the century where flesh-and-blood human beings reassert sovereignty over their lives, their markets and their democracy. . . .

Let us not forget that corporate control was never inevitable. They took it from us, and it is our responsibility to take it back.

Corporations Are Engaging in Phony Environmentalism

by Richard Gilpin and Ali Dale

About the authors: *Richard Gilpin and Ali Dale write for* Red Pepper, *a monthly left-wing journal concerned with protecting the environment.*

Green petrol, ozone-friendly aerosols, plastic nappies that make great compost. . . . If you haven't noticed this kind of oxymoronic labelling the chances are you've been 'greenwashed'. Don't take it personally, you're not alone. The *Concise Oxford English Dictionary* has only recently put a name to one of the most pervasive phenomena of the past 20 years.

Polishing the Corporate Image

Several of the world's largest companies now spend more on cleaning up their image than on promoting their products. The question is: are the public still able to differentiate between change and the illusion of change, or are we too greenwashed up?

The genetically modified (GM) food giant, Monsanto, has been forced to revolutionise its image because of public unease. The seeds of public doubt were sown when the press disclosed Monsanto's plans to introduce a 'terminator' seed that would make crops sterile, forcing farmers to buy fresh seed every year. This would have given Monsanto an almost total monopoly of the seed market in countries like India.

In Britain, [environmental activist group] Greenpeace's involvement in smashing GM crop trials brought the issue home and public pressure forced Monsanto to back down. Chief executive Robert Shapiro announced: 'We are making a public commitment not to commercialise sterile seed technology.' Indian farmers celebrated and the British public was reassured that its protests had been effective; that multinational companies are here to serve us, not to trick us into dependency on their technology.

But, according to [conservation organization] Friends of The Earth, 'This is a

Excerpted from "The Great Greenwash," by Richard Gilpin and Ali Dale, *Red Pepper*, June 2000.

totally empty gesture. They [Monsanto] have had appalling press recently and this is simply an attempt to redress the balance. It hasn't even closed the deal to get the technology for the 'terminator' seed, let alone develop it and they can always introduce it at a later date, when the fuss has died down.'

Unfortunately, greenwashing is neither confined to the corporate boardroom nor to straight exercises in public relations. Governments and so-called 'independent' scientific organisations are equally concerned to get public opinion on-side. We saw this several months ago, when the *Guardian* reported that threats were made to the editor of the medical journal *The Lancet* over his move to publish research questioning the safety of GM food. It was alleged that an influential group within the august [British scientific academy the] Royal Society had set up a 'rebuttal unit' to counter criticism of GM food and push an unequivocal pro-biotech line, something the society rebuffed in the paper's letters pages. Any such efforts to influence debate would be as much of a greenwash as Monsanto's eco-plastic promotions.

Paying Lip Service to the Environment

In the business world, the past 20 years have seen greenwashing evolve from the relatively inane—such as Marks and Spencer's in-store placards urging shoppers to 'Return your trollies [cart]: protect the environment'—into today's sophisticated multi-million pound PR [public relations] campaigns. It was first developed in the mid-1980s, when, prompted by concern over global warming, many global companies created strategies for dealing with public concern. Proctor and Gamble, for example, suddenly announced that its disposable nappies [diapers] were environmentally friendly. To press the point home, each packet carried the World Wildlife Fund (WWF) logo. It later transpired they had paid £300,000 to use the logo without the WWF having validated the company's claims. By the time the Advertising Standards Agency had found Proctor and Gamble's campaign to be 'misleading', the boundaries of greenwashing had moved beyond straight marketing.

> ***"Several of the world's largest companies now spend more on cleaning up their image than on promoting their products."***

In 1993, public pressure forced Dupont, one of the major multinationals, to stop the use of CFCs (chlorofluorocarbons) in Western countries. Its solution was to introduce HFCs (hydrofluorocarbons), which the public understood to be harmless, particularly as they were promoted as 'protecting the skies'. However, all Dupont had actually done was shift CFC production to the Third World while providing the 'developed' nations with HFCs, themselves powerful greenhouse gases. As both CFCs and HFCs are invisible and undetectable, the public had no way of knowing the difference. We'd been greenwashed.

Most multinational corporations now boast 'codes of conduct' as part of their environmental policies. But the resulting self-evaluations, which are not open to public inspection, pay little more than lip service to genuine environmental concerns. Green-sounding corporate front groups are another favoured method. A classic example is the 'Business Council for Sustainable Development'—a coalition of 48 multinational companies that formed in preparation for the 1992 Earth Summit in Rio de Janeiro, Brazil, and which used its considerable influence to weaken key agreements on climate and biodiversity.

Another form of greenwashing is the 'third party technique', where scientific expertise is used to back corporate claims. Often, however, the scientist received funds from the promoters. For example, a professor of environmental science at a US university who has rigorously disputed the evidence suggesting that global warming is the consequence of modern industrial activities received more than $165,000 for research from fuel companies.

Just when you think you're getting to grips with greenwashing, it takes another twist. Companies now shy away from easy-to-spot PR assaults on environmentalists as tree-hugging fanatics putting plants before people. These days, 'co-opting' the green movement is the favoured tactic and establishing a dialogue with 'opposition' groups is seen as a more productive way of overcoming objections to business operations. As James Harris, vice-president of Hill and Knowlton PR, noted: 'For corporations, environmental groups offer the opportunity to obtain positive publicity and gain access to group members. . . . They also provide credibility, which can be particularly valuable.'

"Most multinational corporations . . . pay little more than lip service to genuine environmental concerns."

Information gained from such co-operation can be used to work against and destabilise the group. Usually it is impossible for groups to gauge the impact of their research and information on a company's policies and actions until it is too late to withdraw their co-operation. Companies may, for example, publicise the fact that they have employed an environmental group to review their operations, but disregard the resulting information.

Deceptive PR Campaigns

'Do not change performance, but change public perception of business performance,' says business consultant S Prakesh Sethi.

An interesting new greenwashing campaign comes from Shell, which recently donated, with great fanfare, $200,000 to environmental groups, and embarked on a massive PR offensive following the adverse publicity of Brent Spar and its operations in Nigeria [The Brent Spar is an oil drilling platform Shell unsuccessfully attempted to sink in the North Atlantic sea.]. Its latest advertising campaign, thought by many to be a response to criticism of its Nigerian operations, speaks of 'a commitment to support fundamental human rights, a commitment to sustainable development, balancing economic progress with environmental care and social responsibility. Because making a living begins with respecting life.'

Shell has said it will incorporate the Universal Declaration of Human Rights into its business principles, but the Ogoni have accused the 'Shell police'—soldiers seconded to the company to defend its sites—of using extreme force. The company has refused to comment. Campaigners believe this silence, and that of the company over the Nigerian government's detention up until 1998 of 19 anti-Shell Ogoni, as an indication that the company is not clean and green.

"Companies may . . . publicise the fact that they employed an environmental group to review their operations, but disregard the resulting information."

The company has adopted a new 'stakeholder' process whereby interested parties, including environmentalists and indigenous groups, are brought in to discuss planned operations. Mark Moody-Stuart, Shell chairman, has described this as a way to build 'long-term partnership for the benefit of everyone involved'.

The company has been pioneering the new process in Peru where it is exploring for oil in some of the most ecologically sensitive rainforest left on the globe. What was never on the agenda, however, was whether Shell's operations should go ahead in the first place. Instead, the process has divided different groups over whether to participate or not, leaving Shell with its plans intact and facing a less unified opposition. For environmentalists, co-option may turn out to be as strategically hazardous as direct action has been personally dangerous.

'Good PR is the only way to do business these days,' said the Shell press office. Oil spills caused by either sabotage or neglect are reported at least twice a week in Nigeria. Hundreds of hectares of forest are dead. Pools of crude oil stretch in every direction. Several thousand people have died over recent years through oil fires and pipeline ruptures. Yet, last year, Shell spent £32 million on cleaning up its image as opposed to cleaning up the mess it has made in Nigeria. Its new catchphrase, 'Profits and principles, is there a choice?' says it all. The next time you see an advert suggesting unleaded fuel helps remove pollutants from the air, or that 'it was necessary to level this forest in order to save it', it would be pertinent to ask: Does any of this really wash?

Oil Companies Are Harmful Polluters

by Essential Action and Global Exchange

About the authors: *Essential Action is a project of Essential Information, an organization founded in 1982 by former presidential candidate Ralph Nader. Essential Action provides the public with information that it believes is ignored by the mainstream media and politicians. Global Exchange is a liberal research center promoting global awareness.*

There is a long and terrible record of environmental destruction and human rights violations in the oil-producing regions of Nigeria. The gross level of environmental degradation caused by oil exploration and extraction in the Niger Delta has gone unchecked for the past 30 years. Evidence shows that the oil companies operating in Nigeria have not only disregarded their responsibility towards the environment but have acted in complicity with the military's repression of Nigerian citizens. The profit-driven collusion between multinational oil companies and the past and present Nigerian governments has cost many lives and continues to threaten the stability of the region.

Menacing Multinationals

The authors of this report spent ten days in the Niger Delta (September 8–18, 1999) visiting communities that have been affected by the operations of the following multinational corporations: Shell, Mobil, Agip and Elf. Plans to visit areas in Delta State near Chevron Corporation facilities were canceled due to the instability in those areas. However, while in Nigeria, we interviewed individuals who gave personal accounts alleging Chevron's involvement in recent killings in the Delta. We also met with a group of U.S. lawyers who were in Nigeria at the time gathering information to substantiate lawsuits against Chevron in U.S. courts.

During our visits, we met with community residents, leaders of community groups, and state and local government officials. Despite efforts to arrange meetings with representatives of the oil companies, we were only granted one meeting with a representative of Shell. Based on the testimonies of those we met, as well as on our own observations, we conclude that oil extraction and the related operations of multinational oil corporations pose a serious threat to the livelihood of the people of the Niger Delta.

> ***"The gross level of environmental degradation caused by oil exploration and extraction in the Niger Delta has gone unchecked for . . . 30 years."***

Tensions in the Niger Delta continue to erupt into violence as natural resources vital to local communities' survival are destroyed by oil operations. Environmental and social justice for the people of the Niger Delta remain central issues for achieving peace in the region. As long as people's calls for justice continue to be ignored and resisted by both multinational oil corporations and the Nigerian government, the situation in the Delta can only deteriorate. Many ethnic groups in the Niger Delta have produced declarations and bills of rights that call for autonomy in the management and control of local natural resources. We believe that the survival of a large number of Niger Delta communities is now dependent on their ability to establish their entitlement to local resources.

Nigeria, the most populous country in Africa, is also one of the best endowed in terms of natural resources. Yet, it is one of the poorest countries in the world. As is the case with many oil-rich developing countries, oil reserves have proved a mixed blessing for Nigeria. Since 1974, only 14 years after independence, oil production for export has been by far the main source of revenue for the government. Today, oil sales account for more than 40 percent of gross domestic product (GDP), 80 percent of the government's budgetary revenue, and more than 95 percent of exports. With an average production of approximately 2 million barrels per day, Nigeria is one of the world's largest oil producers. However, due to a persistent fall in oil prices, Nigeria's external debt has risen to an unprecedented level in the last decade; inflation is rampant, and per capita gross national product (GNP) has fallen to levels comparable to or lower than those estimated in the mid 1960s, when oil exploration began in earnest.

Few Benefit from Pollution and Repression

The oil industry has expanded in Nigeria at the expense of other previously important production sectors, such as agriculture and manufacturing. This has created regional imbalances and an increasingly unequal distribution of wealth between different sectors of society, deepening the potential for conflict in this complex multi-ethnic nation.

The Niger Delta, one of the world's largest wetlands, and the site of most of Nigeria's biodiversity, is also the area where the main oil reserves are found. Almost one third of Nigeria's oil is shipped directly to the U.S. Most of the balance is sent to other countries, mainly in Europe, and very little remains in Nigeria for refinement and consumption. During the last four decades, hundreds of billions worth of crude oil have been extracted from the Niger Delta wetlands, earning huge profits for a privileged few, while virtually robbing the affected communities of both life and livelihood.

In addition to a clear lack of access to this locally produced resource, the inhabitants of the Niger Delta region have seen few benefits from the large-scale operations carried out in the proximity of their communities. In fact, in their comprehensive 3-year long study, the Human Rights Watch organization states that "Despite the vast wealth produced from the oil found under the Delta, the region remains poorer than the national average; and [. . .] the divisions between the rich and poor are more obvious in the areas where gas flares light up the night sky."

Though oil companies claim that their operations are carried out according to the highest environmental standards, it is indisputable that they have had a severe impact on the environment, and on agricultural and fish production throughout the Niger Delta region. Many communities report they rarely receive any or sufficient compensation for land taken by oil companies, or rendered useless by oil spills, acid rain, and other forms of pollution. Moreover, protests against environmental degradation and loss of land rights by local communities have frequently met with violent repression by the various police and security bodies with the complicity of the oil companies.

"Hundreds of billions worth of crude oil have been extracted . . . robbing the affected communities of both life and livelihood."

The main multinational oil companies operating in the region are Shell (accounting for more than 40 percent of the volume of production), Mobil and Chevron, in that order. Other companies with significant presence in the Delta are the Italian company Agip, France's Elf-Aquitaine (commonly known as Elf), and Texaco. All of these companies operate on the basis of a joint venture with the Nigerian government. . . .

Immediate Effects of Pollution

Gas flaring. Testimonies to wasteful oil industry operations, gas flares are a distinctive feature of the Niger Delta landscape. Most of these flares burn 24 hours a day and some have been doing so for over 40 years. Communities near these flares are deprived of even the comfort of night's natural darkness.

Natural gas is a by-product of oil extraction; it is removed from the earth's crust along with the crude oil. Natural gas does not have to be flared off, and in many countries there is little flaring. Other options for managing natural gas include reinjection into the subsoil, storage for use as a source of energy by local communities, and transportation for use in other projects elsewhere. Yet companies in the Delta opt for flaring because, even with the minimal fine per barrel of gas burned that has to be paid to the government, it is by far cheaper than the alternatives. Though these "savings" may appear rational to companies, the reality is that local communities are being forced to pay the very high cost of losing a potential valuable resource, and living with the resulting pollution.

Though it hasn't been fully assessed, the impact of gas flares on the local ecology and climate, as well as people's health and property, is evident. The extremely high levels of carbon dioxide (CO_2) and methane gases that are released to the atmosphere also impact climate patterns beyond the local level.

"High levels of . . . CO_2 and methane gases that are released [by industry operations] . . . impact climate patterns beyond the local level."

We witnessed many such flares in our visits to communities: their heat was so intense it was impossible to get near them. A constant loud roar accompanied the thick column of smoke emanating from them, fouling the air. The associated gases could be smelled from hundreds of meters away.

Yet, the oil industry seems blatantly oblivious to the consequences of this wasteful practice. We met with Mr. Bobo Brown, Shell Nigeria's Eastern Division public relations officer, who denied that communities were harmed by gas flare pollution, and even claimed that local residents benefited from these flares because they could dry their foodstuffs for free by setting them near the burning gases, a visibly ridiculous cost-benefit estimate.

Acid rain. Acid rain, a direct result of gas flaring, is taking its toll on the Niger Delta. Acid rain not only deprives people of drinkable rainwater and stunts crop growth (as we found in Eket and other communities in Akwa Ibom State), it is also affecting people's homes. In Iko, Eket, and Etagberi we were told that zinc roofs, which formerly lasted 7–10 years (and were a good alternative to labor-intensive thatched roofing), are now destroyed within one or two years by acid rain. This has led many home owners to resort to asbestos roofing, which although is more resistant to acid rain, it is also more expensive and hazardous to health.

Pipeline leaks. In addition to the grave problems associated with gas flares, on-site oil leaks and ruptured pipelines are a serious problem in the Niger Delta. Decrepit pipelines, some reportedly over 40 years old, criss-cross villages and land, some of them above the ground. These pipes are rusty and in obvious need of repair.

On average, three major oil spills in the Niger Delta are recorded each month. In the first quarter of 1997 alone, Shell recorded 35 incidents of oil spills in its operations. In June 1998, it was reported that a leak near the Otuegwe 1 community that had been going on for months had spilled over 800,000 barrels of crude from a 16-inch buried pipeline belonging to Shell. The resulting ecological devastation seriously impacted the residents of Otuegwe 1 community. Villages in many areas claimed that when pipelines corrode and leak, oil workers will inspect but not repair the leak. Instead, villagers say, oil companies often claim sabotage.

Under Nigerian law, companies are not obliged to clean up or compensate for the effects of spills caused by sabotage.

On September 17, 1999 there was an explosion at the fishing and farming community of Ekakpamre, in Delta State. Residents in the Etche area told us about a recent spill that went untended for weeks, even though, villagers said, Shell had been alerted as soon as the leak was discovered. According to the secretary of the local Community Development Committee, Shell's workers repaired the leak the day before the arrival of our delegation to Etche. In describing Shell's reported sluggishness to repair leaks, Chief Thankgod Albert of the Etagberi village, where Shell has 44 wells, said: "They [Shell] don't treat us like humans. They treat us like animals."

The threat of pipeline explosions puts people at risk of death or injury. In October 1998, a pipeline leak that flooded a large region near the village of Jesse exploded, causing the death of over 700 people, mostly women and children. In Ogoni, Rivers State, we saw above-ground pipelines that crossed right in front of people's homes. In the community of Umuechem, Rivers State, we saw above-ground rusty pipelines that stretched as far as the eye could see. Some of these pipes appear to be greatly corroded, which increases the risk of spills.

Long-Term Effects of Pollution

The delegation has reason to suspect that serious respiratory problems witnessed in many communities can be linked to environmental pollution. Respiratory problems, coughing up blood, skin rashes, tumors, gastrointestinal problems, different forms of cancer, and malnourishment, were commonly reported ailments in many communities. Many children have distended bellies and light hair, which are evidence of kwashiorkor, a protein-deficiency syndrome. Residents repeatedly attributed the spread of kwashiorkor in their communities and the drastic decline in fish catch and agriculture to the pollution of rivers, ponds, sea waters and land by oil industry operations.

"On average, three major oil spills in the Niger Delta are recorded each month."

Another problem facing the people of the Niger Delta is the illicit use of land by oil companies. In the community of Umuebulu, Rivers State, hardly 50 meters away from its perimeter, there is an unlined chemical waste pit belonging to Shell. The company reportedly acquired this land under the pretense of building a "life camp"—Shell's lingo for an employee housing complex. We were stunned to see this site through a chainlink fence in the concrete wall surrounding the facility. The wall keeps people out but doesn't serve as a protection against the noxious fumes coming from the site. Some members of our delegation who live near similar waste sites in the United States immediately recognized the smell of industrial waste. The community said that requests for disclosure of information about the source of the smells and their possible effects on health, as well as compensation for already visible symptoms (such as skin rashes) attributed to pollution, have gone unheeded by Shell.

> There was an oil spillage that occurred in Epubu community that was discovered and reported on the 5th and 14th of December, 1998. The operators of the current burst [pipe] . . . is Nigerian Agip Oil Company. And up till this moment of this interview that spill has not been cleaned. The flora and fauna and the entire ecosystem of the place is destroyed. To be candid, I don't know what Epubu community has done to Nigerian Agip Oil Company. We are contributing to the growth of Nigerian Agip oil company. We know that the operators of the . . . oil [companies] are there to maximize [their] profit. But you don't maximize your profit to the detriment of the people. [This] oil spillage that has occurred since December 1998 to 9th of September 1999 has not been cleaned. The government of the state is also aware of that. You can see the level of injustice the community is going through. We have approached Nigerian Agip Oil Company on several occasions to go and clear this spill. We have written [a] series of letters guaranteeing the security of their personnel. Yet Nigerian Agip Oil Company has refused and the ecosystem of the place is destroyed.*

In Umuakuru, Rivers State, we heard of a similar example of misleading practices by the same company. Residents told us Shell had approached the community to obtain approval to build a recycling plant near the village. The community agreed, and the site was fenced all around; nothing else happened for several years. An independent environmental impact assessment commissioned by residents of Umuakuru later revealed that Shell intended to build an incinerator and a composting unit to process sewage sludge for industrial and medical waste from its employee hospital in Port Harcourt. Despite the community's efforts to halt the process, community residents fear the construction will proceed.

The Niger Delta has the third largest mangrove forest in the world, and the largest in Africa. Mangrove forests are important for sustaining local communities because of the ecological functions they perform and the many essential resources they provide including soil stability, medicines, healthy fisheries, wood for fuel and shelter, tannins and dyes, and critical wildlife habitats. Oil spills are contaminating, degrading, and destroying mangrove forests. Endangered species—including the Delta elephant, the white-crested monkey, the river hippopotamus, and crocodiles—are increasingly threatened by oil exploitation.

The construction of infrastructure for oil facilities is done with little or no regard for environmental considerations. To facilitate road construction, waterways are frequently diverted, to the detriment of fish populations.

*Excerpts from the Essential Action/Global Exchange researchers' interview with His Royal Highness, Chief Nikuman Ebe Obom, the Paramount ruler of Epubu, Sept. 9, 1999, Port Harcourt.

Sudden and drastic changes to the local environment by oil companies are sometimes accompanied by direct loss of human life. For example, the Egi community has reportedly lost five children in the last few years who during the rainy season drowned in "burrow pits" dug by Elf to extract sand and gravel for road construction. . . .

The Disturbing Reality

While the story told to consumers of Nigerian crude in the United States and the European Union—via ad campaigns and other public relations efforts—is that oil companies are a positive force in Nigeria, providing much needed economic development resources, the reality that confronted our delegation was quite the opposite. Our delegates observed almost every large multinational oil company operating in the Niger Delta employing inadequate environmental standards, public health standards, human rights standards, and relations with affected communities. These corporations' acts of charity and development are slaps in the face of those they claim to be helping. Far from being a positive force, these oil companies act as a destabilizing force, pitting one community against another, and acting as a catalyst—together with the military with whom they work closely—to some of the violence racking the region today.

Nigeria is the world's 13th largest oil producer, yet it was until recently chronically short of fuel, having to import it from other oil-producing nations. Though the government is a 55–60% shareholder in oil operations and earns billions in royalties each year, local infrastructure at the source of these billions is in shambles, food shortages abound, malnutrition is common among Niger Delta children, power blackouts regularly occur, and roads are usually in terrible condition.

Everywhere we visited we witnessed the destruction of the local environment, and the oppression of communities affected by what can accurately be described as an outlaw oil industry. Under the somber shadow of this industry of wealth, millions of Niger Delta residents try to survive. The tragedy of so much oil being extracted from the same lands where abject poverty has become institutionalized is unbearable. Over the last 40 years, billions of dollars in profits are earned each year, as millions of barrels of oil are extracted. Meanwhile, high unemployment, failing crops, declining wild fisheries, poisoned waters, dying forests and vanishing wildlife are draining the very life blood of the region. Even the rainwater is acidic and poisoned. What else can the oil companies take from the people? And, what should they be required to give back?

It is a sad reality that Nigeria's oil helps fuel the industrialized world in its mad rush for "progress," while the producing nation is left so obviously far behind. Nigeria still needs to recover the nearly $55 billion in oil profits stolen by the military rulers over the last 15 years. Debt relief and poverty alleviation programs are also desperately needed. The Nigerian human rights community, which includes many of the brave non-governmental organizations (NGOs) and community leaders whom we met, needs governmental protection, not persecution. An open and honest dialogue is called for between the leaders of the oil-producing communities and the oil companies towards resolution of the crisis that meets the needs of both residents and producers. These corporations must adhere to the minimum operational criteria that exist within their own home nations.

Businesses Have Reduced Pollution

by Lynn Scarlett and Jane S. Shaw

About the authors: *Lynn Scarlett is the executive director of the Reason Public Policy Institute, a conservative think tank. Jane S. Shaw is a senior associate with the Political Economy Research Center (PERC), an organization interested in free-market solutions to environmental problems.*

Just as people sometimes litter, businesses sometimes pollute. Just as some people are reckless, some businesses are wasteful. However, over time, the environmental record of business is good.

- Businesses have become cleaner.
- They use fewer materials per unit of production.
- They create the technology that enables consumers and industry to conserve resources, minimize pollution, and enhance their surroundings.

Many of these accomplishments have occurred as a natural consequence of a competitive marketplace. Others have occurred as the result of deliberate efforts to reduce pollution or enhance the environment—either in response to regulations or to market forces.

Prices and Profits: Friends of Conservation

The private sector's record in conserving natural resources is impressive. Even though industry is often accused of using up natural resources for sometimes frivolous purposes, in fact the efficient use of resources is a hallmark of business.

Consider the aluminum beverage can, a familiar everyday container produced by the billions each year. According to [researchers] William F. Hosford and John L. Duncan, manufacturers of aluminum cans "exercise the same attention and precision as do makers of the metal in an aircraft wing." Years of research and development have made the walls of cans thinner than two magazine pages yet able to withstand more than 90 pounds of pressure per square inch.

This improvement is more than an engineer's delight. It is testimony to the

Excerpted from "Environmental Progress: What Every Executive Should Know," by Lynn Scarlett and Jane S. Shaw, *PERC Policy Series*, April 1999. Reprinted with permission.

ability of industry to reduce its use of materials. Today's aluminum cans are at least 27 percent lighter than they were in the 1960s, and they are 80 percent lighter than state-of-the-art steel cans of three decades ago. And the search for material reduction continues. Every additional 1 percent reduction in aluminum that the industry can wring out will save $20 million or so in metal costs.

Aluminum is not an exception. The history of beverage packaging is a history of resource reduction—using less material per unit of output. Aluminum's success forced producers of other materials to become more efficient as well. Steel, the first metal for beverage cans, dropped out of the beverage-can race, but plastic and glass manufacturers improved their use of resources. Instead of heavy bottles and cans, we have lightweight ones, whether of aluminum, glass, or plastic.

Beyond packaging, the profit motive has led to reductions in material use in industries from forest products to automobiles.

- Steel high-rise buildings today require 35 percent less steel than the same building would have required two decades ago.
- A fiber-optic cable made from 60 pounds of sand can carry many times more information than a cable made from 2,000 pounds of copper.
- Improvements in lumber production have reduced wood losses from 26 percent in 1970 to less than 2 percent in 1993.

Surprising Reductions in Pollution

In the late 1970s, two economists surveyed all the data they could find indicating long-term environmental trends. They were surprised. Concentrating primarily on the period between 1940 and 1975, they found that "the trends in environmental quality run the gamut from steady deterioration to spectacular improvement." They had expected that "virtually all forms of environmental damage were increasing and that, in the absence of powerful countermeasures, they would continue to accelerate more or less steadily."

One reason for the improvement is the often-maligned profit motive, which has actually placed a steady downward pressure on pollution. For example, smoke is a sign of poor combustion, indicating that fuel is being lost to the atmosphere through the smokestack. Throughout history, competition has forced firms to minimize all costs, including the cost of unburned fuel. Over time, smoky chimneys gave way to better combustion or cleaner fuels. Emissions of air pollutants declined.

> ***"The often-maligned profit motive . . . has actually placed a steady downward pressure on pollution."***

Indeed, a number of analysts have studied the history of air pollution. They discovered that emissions of some critical air pollutants were going down significantly well before passage of the federal Clean Air Act.

Other factors also have spurred business to reduce pollution:

- *The common-law tradition.* Polluters are liable for the harm they cause. Individuals and owners who are harmed can sue in court for relief. If they can show the cause of the pollution, courts may require that damages be paid or enjoin the pollution. We have a long history of such protection.

• *Protection of reputation.* Companies want to preserve their reputations. Their "brand name capital" is at stake if customers perceive them as polluters.

• *Public pressure.* The public has pushed successfully for laws and regulations to reduce pollution. Regulation of pollution on a local basis in the United States goes back to colonial times.

• *Changing attitudes.* Businesses are situated in a larger social culture whose values are gradually but constantly changing. This change reflects, in part, higher incomes. When requirements for basic food and shelter absorb the attention of most of humankind, many environmental values are neglected. But as basic survival needs for most people have been met in the United States, people have become more willing to devote income to greater environmental protection.

Business practices reflect these changing values. In the late nineteenth century, the focus of business was on mass-producing goods that could be sold cheaply to large numbers of people. Through most of the twentieth century, the focus was on streamlining production, fine-tuning product quality, and tailoring products to specific consumers. By the late twentieth century, as people were placing more value on the quality of life, businesses began to emphasize reducing environmental impacts from manufacturing processes and products. Many businesses now routinely incorporate environmental considerations into their process and product design. This emphasis is sometimes termed "industrial ecology."

"Many businesses now routinely incorporate environmental considerations into their process and product design."

Some businesses are even developing their central business strategy around environmental goals Increasingly, financial institutions track environmental risk. According to *Business Wire*, one firm, Innovest Capital Risk Advisors, evaluates firms by reviewing their historical liabilities, operating risks, and potential exposure to risk and relates this environmental risk performance to the firms' share-price performance. Their findings show that "ecoefficiency"—how well companies manage environmental risks and exposures—is a strong indicator of superior corporate management, superior financial performance, and shareholder value.

Whether business executives embrace the new environmental values or simply resign themselves to them, they recognize the existence of these values and the pressures they exert. Companies that deal effectively with these pressures are likely to do well and those that do not will falter.

One reason that some people criticize business on environmental grounds is that they have absorbed the popular view that corporate malfeasance is widespread and to be expected when regulation is weak. The story of Love Canal epitomizes the popular view of business.

Indeed, the story sounds bad for business: A company used an old canal bed in the city of Niagara Falls, New York, for a waste dump. Twenty-five years later chemicals leaked out, injuring children and coloring lawns. This news frightened local residents, created a national chemical scare, and led to the passage of a law that taxed chemical companies to clean up waste dumps—a 1980 law now known as Superfund. The law was designed to stop the "ticking time bombs" around the country.

"Some people . . . have absorbed the popular view that corporate malfeasance is widespread and to be expected."

Yet seldom has an environmental problem been so misunderstood. Eric Zuesse, writing in *Reason* magazine in 1981, discovered by reading through old records that it was the Niagara Falls Board of Education, not Hooker Chemical Company, that had acted irresponsibly.

Love Canal was an empty canal bed that Hooker bought in the 1940s for disposal of chemical wastes. After preparing the site and using it for some years, the company sealed the dump with clay. For the most part, it used state-of-the-art techniques that were as stringent, as it turned out, as national regulations in effect in 1980.

In 1952, the Board of Education of Niagara Falls, looking for land on which to build a school, pressured the company, under threat of condemnation, to sell land that included the dump. Hooker ultimately sold it to the board for one dollar. Hooker specified in the deed that the land had been filled with chemical waste products, and Hooker officials escorted school board officials to the property and made test borings in their presence to confirm the presence of chemicals. In addition, the company tried to include a clause in the deed that would assure that the land over the dump would be used only for a park, but the board refused that limitation.

As time passed, the school board sold off some of the land to a developer—exactly what Hooker had warned against. Most importantly, the school board allowed sewers to be built that broke through the protective clay wall. Later, construction through part of the property also exposed chemicals, and, ultimately, chemicals seeped into people's homes, bringing about the famous Love Canal scandal.

Constrained by liability, as private companies and private citizens are, Hooker for the most part acted properly. It covered the dump and informed the buyer of its potential danger. But the Board of Education disregarded the dangers. As public officials, the board members were not personally liable for their actions. They knew of the dangerous conditions before selling, but were too concerned about providing education at a reasonable cost to heed the warnings.

A single example such as Love Canal cannot justify general conclusions.

However, the case is instructive. Not only does it force us to question some "conventional wisdom" about environmental matters, it also sheds light on the past and how environmental problems were addressed in the past.

The case shows that liability, enforced through the courts under common-law rules, constrains private-sector actions and that Hooker Chemical responded to these liability constraints. Yes, it placed hazardous chemicals in the ground, but contained them on its own property and took steps to keep the land from being used in ways that might lead to harm. Ultimately, in federal court, Hooker was held negligent for not having done enough to warn the school board, though the court recognized that Hooker had attempted to limit use of the contaminated land.

The Love Canal case shows that historically companies, constrained by legal precedent, acted to avoid causing harm from pollution. The resulting protection extended to individuals of modest means and small power in the community. While it was not perfect, it prevented or corrected a great deal of environmental harm.

Ironically, given the attention paid to Love Canal and other hazardous waste sites, the contamination from Love Canal is small compared to the contamination now acknowledged to exist at Department of Defense and Department of Energy facilities. For example, the final cleanup tab for all Superfund and military sites in Colorado alone may exceed $12 billion, or $46 for each U.S. citizen. Of that figure, $11.8 billion will go toward cleaning up two government sites. Only a small fraction of the costs will be for "orphaned" sites in which the former private owners are now bankrupt or cannot be located—even though that was the original purpose of the Superfund legislation. Colorado is not unique. The *New York Times* reported that hazardous waste per acre was twice as high on federal land as on private land.

U.S. Factories in Mexico Are Reducing Pollution

by the Environmental Protection Agency

About the author: *The Environmental Protection Agency (EPA), established in 1970 by then-president Richard Nixon, is the government agency in charge of implementing and enforcing pollution regulations.*

Several large U.S. companies are taking the lead in pollution prevention on the Mexico side of the U.S.-Mexico border. Other companies are following the lead because it's good business.

Dr. Pierre Lichaa, border pollution prevention coordinator for the Texas Natural Resource Conservation Commission, says American companies that operate on the Mexico side of the border are cutting pollution and also saving money by integrating environmental management into their business goals.

It hasn't always been that way.

A Booming Border

Border pollution began to increase after 1965 when Mexico started offering incentives for product assembly plants called maquiladoras, Mexican subsidiaries of foreign-owned companies, mostly U.S. and mostly located near the U.S. border to exploit American markets.

The idea caught on and by 1990 over 2,000 plants were operating. Today, more than 4,000 maquiladoras employ a million people. Within 60 miles of the border 11 million now live in four U.S. and six Mexican states, and the population is expected to double by 2020.

By the mid 1990s, Texas and industry were making good progress at preventing pollution along the U.S. side of the border using compliance tools like on-site assistance, workshops, guidance publications, a Web site, and a hotline. So maybe the good being done in Texas could also be done in Mexico, where pollution concerns were growing as fast as industry.

Mexican officials were willing to try a similar program, and soon Texas spe-

From "Profitable Pollution Prevention," by the Environmental Protection Agency, www.epa.gov, June 13, 2001.

cialists were visiting companies in Mexico to see how they could help, to find out what the companies needed, what they would like to do, and how they would like to do it.

At the same time, the Environmental Protection Agency (EPA) and its Mexican counterpart supported industry regulatory workshops to promote environmental auditing, a strategy that promotes pollution prevention through voluntary compliance, a smart idea since it can save companies money.

The two partner agencies also attacked critical problems like massive pollution generated from over 200 open-burning brick kilns in Juárez. EPA helped with research and technical assistance, and soon Mexico opened a brickmakers' training center in Juárez, and the number of kilns using tires, plastics, and other high polluting fuels dropped significantly.

"American companies that operate on the Mexico side of the border are cutting pollution and . . . integrating environmental management into their business goals."

Once a "bridge of trust" was built through such cooperative efforts, and through workshops and company visits, Dr. Lichaa says many maquiladoras began to set up permanent pollution prevention programs. A team approach brought local university experts and graduate-student engineers into the planning process, and the bridge of trust across the border now linked local companies with local environmental specialists. Though still available for consultation, the Texas partners could move on to other companies.

Success Stories

Dr. Lichaa says several maquiladoras have made excellent progress in the past few years, including Delphi, Alcoa, Kemet, Johnson Controls, and TRW (Thompson, Ramo, Wooldridge). They are reducing air and water pollution and hazardous waste while conserving water, recycling, and becoming involved in local conservation programs. At the same time, they are saving millions of company dollars.

Delphi Rimir in Matamoros, where 850 workers make bumpers for U.S. cars, not only reduced chemical waste significantly, but implemented an ambitious recycling program. By treating and recycling wastewater for irrigation, fire fighting, and sanitary systems, the company no longer discharges water into the Matamoros sewer system. Company savings in materials and reduced cleanup efforts amount to $1.7 million a year.

In Chihuahua, the 1,700 TRW employees that manufacture auto air bags eliminated tons of toxic waste, and they now reuse 100 percent of nonhazardous scrap materials. Working with the local community, TRW builds wheelchairs from scrap metal and air bag materials for less than a tenth of the cost of wheelchairs on the local market.

Using an outside contractor, TRW built small and very affordable model housing units and a 6,000 square-foot daycare center for the children of its employees from the facility's general scraps. And by preventing pollution in such a beneficial manner, TRW saves itself $1.3 million a year.

Other maquilas are recycling used oil, reusing water for things like roof-top cooling systems, using water-based cleaners instead of toxic cleaners, using more efficient light fixtures, reducing use of steel barrels, using recyclable cardboard pallets instead of wood, and recycling other non-hazardous waste.

Of course, these are only a few of thousands of maquiladoras along the border, but these are large companies with multiple plant sites, and they are leading the way in pollution prevention. Each year more companies volunteer for environmental audits, and in 2000 470 of them were awarded Mexico's "Clean Industry" status, including those companies mentioned above. Dr. Lichaa believes many other firms will follow when they understand the health benefits for their communities and the savings for their companies.

Oil Companies Are Protecting the Environment

by the American Petroleum Institute

About the author: *The American Petroleum Institute (API) is the primary trade association representing the oil and natural gas industry in the United States.*

Good air quality is important to the oil and natural gas industry just as it is to all Americans. We want the benefits that petroleum products bring to make our lives better but not at the expense of cleaner air. Fortunately, Environmental Protection Agency (EPA) statistics show that Americans are enjoying steadily improving air quality. "Criteria" pollutant emissions have been reduced, and pollution concentrations measured in the air are lower. Moreover, the lion's share of this progress is attributable to cleaner cars, fuels, and industry facilities and operations. Together, they account for approximately 70 percent of total emission reductions nationwide since 1970. . . .

Cutting Emissions at Industry Facilities

Fuels are cleaner, and so are oil and gas industry facilities. Improved equipment at these facilities, better training and innovative technology, including, for example, advanced computer software, which can both reduce operational costs and help ensure a higher level of environmental performance, have meant better air quality for our employees and our neighbors. Our facilities also provide good jobs, which strengthen local businesses and make it easier for everyone to take better care of their families.

As a result of these advances, refineries and exploration and production operations have substantially cut their criteria pollutant emissions. Between 1970 and 1998, they fell by 69 percent.

Storage and transport operations, including service stations and distribution terminals, are also cleaner. For example, many service stations now have specially equipped pumps that reduce the loss of fuel vapors during fill-ups. Between 1985 (the first year EPA statistics are available for this category) and 1998, criteria pol-

Excerpted from "Overview of the Oil and Natural Gas Industry's Environmental, Health, and Safety Performance," by the American Petroleum Institute, *Protecting Our Environment Report*.

lutant emissions from storage and transport declined by 20 percent.

Under EPA's Toxics Release Inventory program, releases from refineries, which consist mostly of air emissions, have followed the same pattern. Between 1988 and 1997, releases of these emissions fell by 41 percent. Emissions classified as carcinogens, such as benzene, dropped 51 percent.

Currently, EPA data show that refineries contribute less than 3 percent of all of the nation's toxics releases.

Energy and Climate Change

Although global climate theories remain a subject of debate, oil and natural gas companies are closely managing energy use in ways that reduce or control the

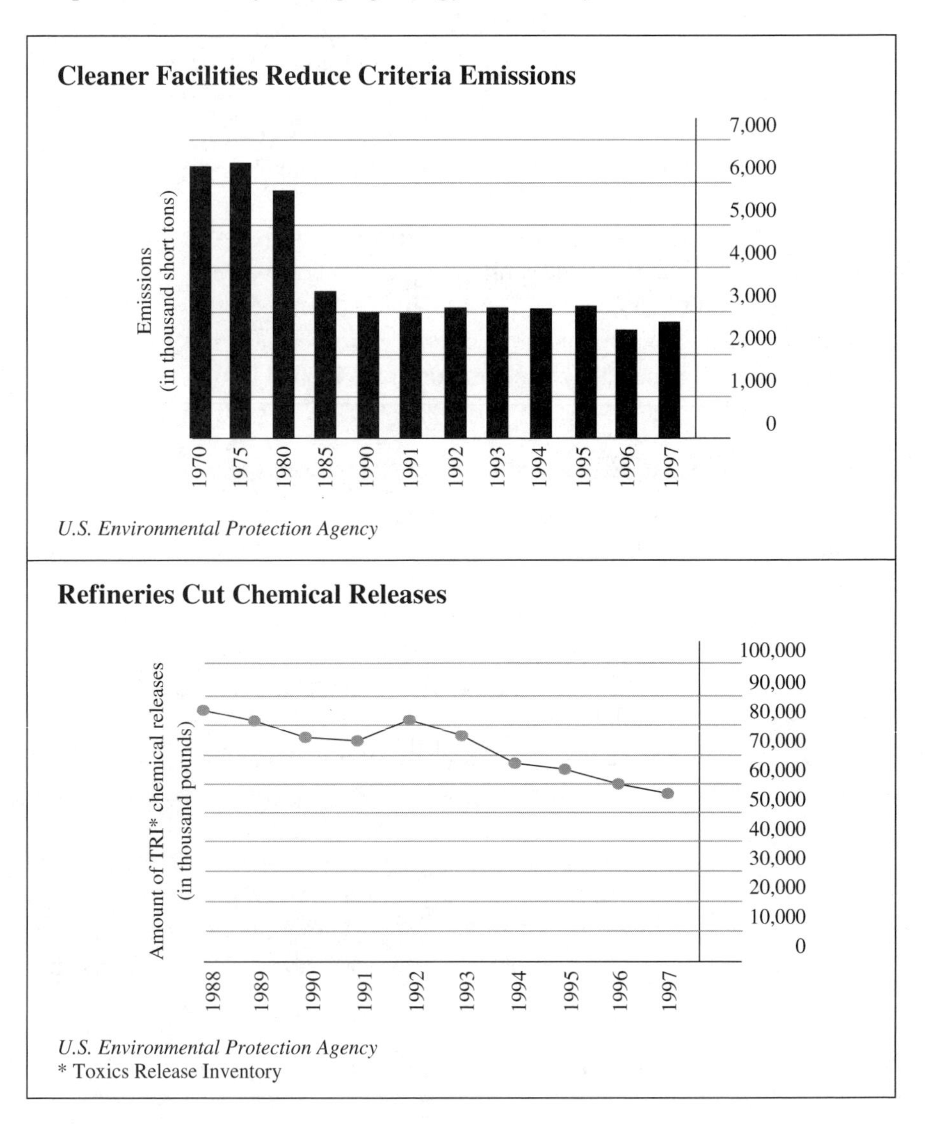

growth of carbon dioxide and other greenhouse emissions. These efforts also cut emissions that contribute to urban air quality problems, conserve resources, and lower energy costs, which are a substantial expense at most industry facilities.

Improvements in energy efficiency include more efficient lighting, equipment, production processes, energy generation, and facilities to capture and reuse waste energy. For example, by switching to energy-saving lighting, one oil and gas company is saving up to 800 pounds of carbon dioxide emissions per year at each of its retail outlets.

Many oil and gas companies are participating in EPA's *Green Lights, Energy Star* and *Natural Gas Star* voluntary programs to increase energy efficiency and reduce greenhouse gas emissions. The industry is also developing a common methodology for measuring and reporting greenhouse gas emissions.

A new report, *Voluntary Actions by the Oil and Gas Industry: A Conference on Industry Best Practices to Improve Energy Efficiency and to Reduce Greenhouse Gas Emissions* demonstrates that oil and gas companies from around the world are not waiting for all the answers before taking action. . . .

Protecting Our Water

Clean water is a vital resource, which the oil and natural gas industry has been working to protect in many different ways—from improving the designs of its ocean-going tankers to upgrading its underground storage tanks. Yet the industry is concerned about more than preventing pollution. It also seeks to enhance water environments near its operations and even extends usable water supplies by providing treated water from production operations for irrigation and for consumption by wildlife and livestock. For example, one company in California provides millions of gallons of treated water from its facilities to help grow grapes, citrus, almonds and pistachios.

Images of oiled beaches are rare today because much less oil is spilled. U.S. Coast Guard data show that less oil was spilled into U.S. navigable waters in 1998 than in any year since 1980.

In the past five years, 1994–1998, about two-thirds less oil was spilled than in the preceding five years, 1989–1993.

In 1990, Congress unanimously passed the Oil Pollution Act, which established new requirements to help reduce oil spills from tankers and other vessels. Since 1991, there have been no large spills over 5,000 barrels from oil tankers in U.S. waters. . . .

> ***"Refineries and exploration and production operations have substantially cut their . . . pollutant emissions."***

To have gasoline ready for consumers when they need it, it is usually stored in underground storage tanks at service stations and other locations. In light of recent concerns about detection of MTBE (a gasoline additive) in some water supplies, American Petroleum Institute (API) member companies have renewed their support for strict enforcement of new underground tank regulations to help reduce leaks or spills that may be contributing to the problem. Under these regulations, API member companies have spent $1.2 billion upgrading or replacing their tanks.

API is also working with air quality regulators in the Northeast, the American Lung Association, the Natural Resources Defense Council and others in support of legislation to address concerns about MTBE. Our mutual goal is to foster enhanced protection of our water supplies while continuing to provide America with the cleaner-burning gasolines needed to improve air quality.

Nearly half of the world's seaborne trade consists of crude oil or petroleum products, including more than half of the crude oil Americans consume. The tankers that move large volumes of crude oil across the seas are the largest ships ever built, as much as a quarter of a mile long.

To prevent or minimize the size of oil spills, single-hulled tankers are being phased out and replaced with double-hull tankers. Worldwide, more than 700 double-hull tankers have been built and put into service through the 1990s, with more than 500 coming on line in the past four years.

By 2015, all tankers operating in U.S. waters will have double-hulls.

New technology is pushing the frontiers of oil and gas exploration and production. Thanks to such advances as tension-leg production platforms, companies can now produce oil and gas more than a mile under the water. Whereas a conventional platform has a rigid structure anchored to the sea floor, tension-leg platforms float attached to the bottom by steel tendons.

"Since 1991, there have been no large spills over 5,000 barrels from oil tankers in U.S. waters."

Other advanced technologies used in offshore operations include robot-like "remote operators" the size of cars, which are used for cleaning platforms and preparing undersea wells—and "virtual reality" computer models for planning new platforms. These models are 3D images of platforms, which are so sophisticated that an operator can conduct a virtual tour to check clearances and identify hazards before a single component is ever built. Early detection of design problems enhances the efficiency and safety of facilities and reduces costs.

Recycling and Conservation

The oil and natural gas industry promotes recycling and conservation to extend the uses of its products and enhance the environment. Companies are cutting energy consumption, reducing and recycling wastes, using renewable fuels, developing wildlife sites and conservation programs, planting forests, buying back gas guzzlers, and restoring land affected by industry operations.

Out-of-the-box thinking has advanced conservation in unusual ways. For example, one company has created artificial reefs out of discarded auto tires. Another has converted tank sludge into fertilizer used to grow orange, mango and banana trees.

Petroleum companies have been at the forefront of the nation's used motor oil collection programs. API members and their independent dealers now operate over 12,000 drop-off collection centers in the United States, including service stations and quick lubes. They collected more than 80 million gallons of used motor oil between 1991 and 1996. About 60 percent of all the motor oil in this country is recycled, including virtually all that is collected at service bays. Used oil still has lots of value to meet lubricating or energy needs.

The oil and gas industry is also encouraging used motor oil recycling worldwide. For example, one API member company operating in the Pacific region regularly picks up used motor oil in Fiji and the neighboring Cook Islands, recycling it as burner fuel.

The industry also helps educate the public about the importance of used oil recycling and has encouraged states and municipalities to develop their own programs, increasing the convenience for do-it-yourselfers who want to properly dispose of their used motor oil. . . .

"The oil and natural gas industry promotes recycling and conservation to . . . enhance the environment."

Proper disposal of used motor oil is good for the environment—but disposal doesn't mean throwing it away. Refined, used oil can be used repeatedly as motor oil and other lubricants.

Or, it can be reprocessed and burned as fuel. Reprocessed motor oil helps run power plants that produce electricity. It can be used in industrial burners and in marine fuel. It can also be mixed with asphalts.

Two gallons of reprocessed motor oil can generate 36 kilowatt hours of electricity. That's enough to run the average household for a day, cook 48 meals in a microwave, blow hair dry 216 times, or operate a television set for 180 hours. . . .

It's impossible to count the uses—and reuses—of plastics, which are typically made from petrochemicals derived from oil and natural gas. Discarded plastic computer housings are converted to pothole filler and highway underlayment. Used PET mineral water bottles become sweaters. And worn out plastics-based carpeting is reborn as outdoor parking bumpers. By recycling, you are helping to conserve these natural resources for future generations.

Since 1990, the plastics industry—both as individual companies and through organizations such as the American Plastics Council—has invested more than $1 billion to support increased recycling. The number of plastics recycling businesses has tripled since 1990, now totaling more than 1,700.

Environmental Spending

Meeting Americans' needs for oil and gas also requires substantial investments to protect our environment and that of future generations. Between 1990 and 1998, the U.S. petroleum industry spent almost $83 billion dollars on the environment—including about $8.5 billion in 1998. Spending in that year was more than EPA's total budget and more than double the net income of the top 200 oil and natural gas companies.

The refining sector has typically been responsible for half or more of all industry environmental expenditures. This reflects investment over the past decade—and especially during the mid-1990s—in new equipment and facility improvements to make new, cleaner-burning fuels.

Technological Improvements

Companies have also spent large sums on improved technology, not always counted as an environmental outlay, which also helps the environment.

For example, three-dimensional imaging provides better information about the likelihood of finding oil and gas, reducing costs by avoiding unproductive drilling. But less drilling also means a smaller environmental footprint. Three-dimensional imaging uses sound waves that pass through the earth, are reflected back, and then recorded. Using computers geophysicists translate the data into three-dimensional pictures of the subsurface to help show them where the oil and gas are hiding.

The same imaging technology is used to locate horizontal extended-reach wells, which reduce the number of oil and gas wells typically needed because they drain much larger areas underground. Horizontal wells can also be grouped on one site to limit the surface footprint.

Over 150,000 miles of pipelines crisscross the United States quietly carrying each year over 525 billion gallons of crude oil and petroleum products, such as gasoline, heating oil and jet fuel. They constitute the nation's most important mode of petroleum transportation—and one of the safest and most reliable, thanks in part to sophisticated inspections.

> ***"Over the past 30 years, the amount of oil spilled from U.S. pipelines declined by 60 percent."***

For example, in Alaska, aircraft equipped with infrared photography locate problem areas in aboveground pipelines before leaks occur. The same technology has also been put to use surveying caribou and polar bear populations to ensure they are not adversely affected by petroleum operations.

Companies also maintain the safety and integrity of pipelines through corrosion protection systems and community programs to prevent third-party excavation damage as might result from accidents involving farm machinery or construction equipment such as backhoes. Over the past 30 years, the amount of oil spilled from U.S. pipelines declined by 60 percent.

Oil and gas companies are concerned about the safety of their employees as well as the impact of their operations on surrounding communities.

Although work in the petroleum industry is often physically demanding and involves heavy equipment or high temperatures and pressures, the job-related illness and injury rate for the industry's workers is well below that of the U.S. private sector as a whole. Between 1989 and 1998 (the most recent 10-year period with available data), the industry rate declined by 20 percent. The number of job-related fatalities also declined—by 11 percent between 1992 (the year the Bureau of Labor Statistics began its fatality census) and 1998.

Chapter 3

Are Pollution Regulations Effective?

Regulation and the Environment: An Overview

by Mary H. Cooper

About the author: *Mary H. Cooper is a staff writer for the* CQ Researcher, *a weekly report on various issues confronting lawmakers and the public.*

First the good news: "We've really made tremendous progress in the United States since the birth of the environmental movement 29 years ago," says environmentalist Paul Portney, president of Resources for the Future. "The air is dramatically cleaner, big strides have been made in reversing water pollution and there is much more careful treatment of hazardous waste and toxic substances."

More or Less Environmental Regulation?

Now the bad news: According to many scientists and environmental advocates, the successes of the past few decades represent just the first, easy steps in the battle to restore planet Earth to health. The next phase of environmental action, they say, will have to address much tougher threats—among them global warming linked to fossil-fuel combustion, water pollution from multiple sources and air pollution that crosses state and even national borders.

"Over the last 30 years, America has made great progress," says Carol M. Browner, [former] administrator of the U.S. Environmental Protection Agency (EPA). "But the job is not done. Today we face a new, and somewhat more difficult, generation of pollution problems.

"To cite a major example, polluted runoff from our city streets, suburban lawns and rural areas today accounts for more than half of all water pollution nationally. We have taken unprecedented steps to reduce polluted runoff, but it will take all of us—the federal government, states, communities and businesses—to solve the problem."

Business representatives, however, say the improvements in the nation's environmental quality demonstrate that the need for massive federal intervention to reverse environmental damage is now obsolete. They believe that further

Excerpted from "Setting Environmental Priorities," by Mary H. Cooper, *Congressional Quarterly*, May 21, 1999.

progress should rely more heavily on state and local government initiatives and on voluntary efforts by businesses, consumers and other potential polluters.

"We still have the best environment in the world," says William L. Kovacs, vice president for environmental and regulatory affairs at the U.S. Chamber of Commerce. "You have to ask what it is we're trying to do today and whether there are real health and safety issues at stake. If there are, the business community will go along with it. But if there aren't, we won't." . . .

Other critics of current policy agree that economic development—not government regulation—is the key to improving environmental quality. "A healthy environment and a healthy economy are directly linked," writes H. Sterling Burnett, environmental policy analyst at the National Center for Policy Analysis, a conservative think tank. The significant reductions in air and water pollution in the United States over the past three decades, he notes, occurred at a time of rapid economic growth. "Pollution wastes resources. In market economies, as companies become more efficient they pollute less.". . .

Opinion polls suggest that voters continue to place environmental protection high on their list of priorities. More than two-thirds of Americans surveyed in a Gallup Poll said they worried "a great deal" about drinking-water pollution. Other big concerns to more than half the respondents included toxic waste contamination, pollution of rivers, lakes and reservoirs and air pollution.

Despite public support for environmental protection, the prospects for major legislative initiatives in this area are uncertain. Since Republicans took control of both houses of Congress in 1995 [as of fall 2001, Democrats control the U.S. Senate], few new environmental proposals have become law. Congressional opposition to targeted cuts in emissions of carbon dioxide and other gases linked to global warming runs so high that former president Clinton did not even submit the 1997 Kyoto Protocol on global warming for Senate consideration and ratification (*see below*). Indeed, conservative lawmakers have tried repeatedly to roll back existing environmental laws by attaching anti-environmental riders to appropriations bills. . . .

Partisan politics, some experts say, has prevented both sides in the ongoing policy debate from undertaking the kind of sweeping review of environmental priorities necessary to address today's threats to the environment and public health. "There's a noticeable absence of long-range thinking and commitment both in Congress and the executive branch," says Terry Davies, director of the Center for Risk Management at Resources for the Future. "Everyone in authority is thinking 10 days ahead instead of taking the big picture and thinking 10 years ahead. That's not good because the process does need some major, long-range thinking, and it's not getting it."

"Critics of current policy agree that economic development—not government regulation—is the key to improving environmental quality."

High on Davies' environmental wish list is a better way to collect, analyze and disseminate data on the environment. "We don't have enough information to tell us where we are or where the trends are going," he says. "We don't really know whether air quality, and especially water quality, are really improving or not under current law. As for solid waste, the situation is hopeless. We don't even know where it is, much less whether it's getting better or worse." As environmentalists, industry representatives and policy-makers consider the current state of the environment, these are some of the questions they are asking.

Should More Be Done to Slow Global Warming?

In December 1997, representatives of the United States and 175 other countries met in Kyoto, Japan, and agreed to take steps to reduce emissions of carbon dioxide and other "greenhouse" gases believed to be causing a dangerous rise in global temperatures.

For decades, the scientific evidence had mounted suggesting that combustion of coal, oil and—to a lesser extent—natural gas was increasing the level of greenhouse gases, which trap solar heat within the atmosphere. If the emissions continued to mount, scientists warned, the polar ice caps would begin to melt, causing water levels to rise around the globe and flood coastal regions. In addition, infectious diseases would spread and food supplies would be threatened by widespread drought.

"Partisan politics . . . has prevented both sides in the ongoing policy debate from [addressing] . . . today's threats to the environment and public health."

The Kyoto Protocol, which the Clinton administration formally signed on Nov. 12, 1998, sought to avert this global environmental disaster by setting targets and timetables for countries to reduce their carbon emissions. The treaty required the United States and other industrial countries, which account for the vast majority of emissions, to make the biggest reductions, while developing countries were given more time to meet the treaty's conditions. The target set for the United States was a 7 percent reduction in carbon emissions below 1990 levels by 2012. [Editor's note: President George W. Bush has withdrawn U.S. support for the Kyoto Protocol as it is currently written.]

Opposition to the Kyoto Protocol runs high in the U.S. business community, which fears that meeting the treaty's emissions targets will raise energy prices, thereby reducing consumption and cutting into corporate profits. Critics also attack the treaty's mild demands on developing countries. Although historically they have contributed only marginally to the buildup of greenhouse gases, some rapidly developing nations—such as India and China—now have some of the fastest rates of growth in carbon emissions. . . .

Some environmental experts are uncertain about the best way to deal with global warming. . . . "Climate change is the mother of all environmental problems today," says Portney. "I think there's no question that it would cost a lot of money to do something dramatic here, but it's the kind of problem you have to take seriously because of the potential environmental risks involved."

In his view, however, the scientific evidence pointing to an imminent threat from global warming is not yet convincing enough to justify requiring U.S. businesses and consumers to shoulder the enormous cost of meeting the treaty's targets, which he estimates at a minimum of $50 billion a year. "It may be that five or 10 years from now we will become convinced that this is a problem that we would really be willing to spend a lot of money on," he says. . . .

> ***"Waiting to act on pollution will only make things worse, many environmentalists say."***

But waiting to act on pollution will only make things worse, many environmentalists say. "We've got to move forward on reducing the greenhouse pollution that comes from our transportation and electricity-producing sectors," says Greg Wetstone, program director for the Natural Resources Defense Council. "This is an issue we're going to have to grapple with, and the sooner we do it the less expensive it's going to be and the easier it's going to be to deal with successfully."

The most effective ways to reduce carbon emissions, Wetstone says, would be to tighten fuel-efficiency standards for vehicles and close the legal loophole that exempts old, coal-fired utilities from the 1970 Clean Air Act's pollution standards. "We could dramatically reduce our pollution just by moving to these common-sense measures in these two big sectors," he says. "The technology is already there to have better fuel economy for motor vehicles across the board. And we need to eliminate this outrageous loophole for coal-fired plants and move to cleaner power production."

Short of amending the Clean Air Act, Congress could help reduce carbon emissions by providing tax incentives for coal-powered utilities to switch to less-polluting plants fueled by natural gas.

"Whether Congress passes utility deregulation or not, the states are moving in this direction, and we will end up eventually with a deregulated utility industry," says Philip E. Clapp, executive director of the National Environmental Trust, which monitors environmental legislation and regulations. "That's going to mean high competition, and there are significant economic incentives for converting old, coal-fired plants, with their high maintenance and operation costs, to natural gas. If lawmakers don't want to accelerate the natural turnover process through regulations, they can always do it through tax incentives."

Clapp and eight other leaders of major environmental organizations criticized the Clinton administration for not pressing more aggressively for reductions in carbon emissions. In a letter to the former president, they expressed "deep disappointment with the lack of an administration proposal to require significant reductions in global warming pollution. We are particularly frustrated that the administration has not sought meaningful emission reductions from either power plants or passenger vehicles.". . .

Are Americans Willing to Pay for Environmental Protection?

In a highly industrialized country like the United States, maintaining a healthy environment does not come cheap. The Clinton administration asked Congress for $34 billion to fund environmental programs in fiscal 2000, a 5 percent increase over the previous year.

Government expenditures are only part of the picture, however. Industry spends billions more to install chimney scrubbers, waste-water treatment facilities and other technologies to limit pollution in accordance with government mandates. Those costs are passed along to consumers, who pay for a clean environment through higher prices for goods and services.

The Chamber of Commerce calculates that the annual cost of complying with environmental laws has risen from $80 billion in 1985 to $170 billion, or about $1,800 per household. "Business has spent a trillion dollars cleaning the environment," the group says. "Additional progress requires regulations based on common sense and sound science. Unfortunately, federal regulations often fall far short."

To environmental advocates, such statements are a familiar refrain. "Any industry that has to do something to cut its pollution immediately turns around and says that it will make Americans lose jobs, bankrupt America and force consumers to give up essential products," says Clapp. "Americans don't have to drive less in order to cut pollution; Detroit has to produce more fuel-efficient cars, which are actually an advantage to the consumer."

But some environmentalists warn that the costs of environmental progress will rise with the next round of programs. "We've picked off all of the cheap fixes," says Portney of Resources for the Future. He cites programs to reduce urban air pollution and water pollution caused by clearly identifiable sources, such as factories or sewage-treatment plants. "After 30 years of environmental legislation, the marginal cost—the cost of additional improvements—is much bigger than it has ever been."

> ***"67 percent of Americans surveyed . . . agreed that 'protection of the environment should be given priority.'"***

The next phase of air pollution efforts, for example, is likely to deal with more complex, and costly, matters, such as regional, rather than local, pollution. Nitrogen oxides emitted by Midwestern and Eastern coal-fired electric utilities is blown across state boundaries and contributes to dirty air in the Northeast. Pollutants from Los Angeles are largely responsible for the haze over the Grand Canyon, hundreds of miles away. Solving such problems will require far more sweeping steps than installing catalytic converters on cars and "scrubbers" on factory smokestacks.

Portney, perhaps surprisingly, questions the need for immediate action in some cases. "Environmental advocates get jerked out of shape when I say this," Portney says, "but I don't think there are very many serious environmental problems left in the United States. That doesn't mean that we shouldn't do more to clean up the environment. But it does mean that we've got to look a lot more carefully than we did back in 1970, when you could throw a dart at a list of environmental problems and feel confident that whatever you hit was a good candidate for regulation. That means that we need better research to make sure that we're not spending our money on problems that aren't worth the money."

Some environmental advocates say Americans don't want to wait for further scientific proof. "Americans feel very, very strongly about cleaning up air and water," Clapp says. "Those are the two strongest issues, and they are willing to pay for them. The arguments of industry, that environmental protection threatens job security, resonate at times of economic hardship. But we've had nine years of extremely healthy economic conditions, and people just don't believe that we can't have good environmental protection and economic growth at the same time."

Recent opinion polls suggest Clapp is right: 67 percent of Americans surveyed on April 22, 1999, the eve of Earth Day, agreed that "protection of the environment should be given priority, even at the risk of curbing economic growth." Only 28 percent of those responding to the April Gallup Poll thought that "economic growth should be given priority, even if the environment suffers to some extent." A survey conducted for the Sierra Club the same month had similar findings. When asked whether they would rather see $2 billion of the federal budget surplus used to buy and protect open space or for military spending, respondents chose environmental protection over defense spending by a margin of 50 percent to 34 percent.

The Superfund Hazardous Waste Program Is Effective

by the Environmental Protection Agency

About the author: *The Environmental Protection Agency (EPA) is the government agency responsible for implementing and enforcing pollution regulations on industry and individuals.*

At Love Canal [an old canal bed in Niagara Falls, New York], over 21,000 tons of chemical wastes were deposited in a landfill. The landfill closed in 1952, and was then covered over the next year. Over time, a community grew around the abandoned landfill. Under the old scenario of "out of sight–out of mind," that should have been the end of the story.

However, more than two decades later, increasing numbers of Love Canal residents began complaining of health problems, including chronic headaches, respiratory discomforts, and skin ailments. Residents also noticed high incidents of cancer and deafness. The State of New York investigated and found high levels of chemical contaminants in the soil and air—with a high incidence of birth defects and miscarriages in the immediate area around the Love Canal landfill. Former president Jimmy Carter declared a State of Emergency in 1978, and Federal funds were used to permanently relocate 239 families in the first two rows of houses that encircled the landfill area.

But the tragedy did not end. A New York State investigation found "extensive migration of potentially toxic materials outside the immediate canal area." In 1979, 300 additional families in a 10-block area around the site were relocated because of health problems from chemical exposures. In 1980, the Environmental Protection Agency (EPA) announced the results of blood tests that showed chromosome damage in Love Canal residents. Residents were told that this could mean an increased risk of cancer, reproductive problems, and genetic damage. Later that year, President Carter issued a second State of Emergency—providing funding for the permanent relocation of all 900 residents of the Love Canal area.

From "Superfund: 20 Years of Protecting Human Health and the Environment," by the Environmental Protection Agency, www.epa.gov, 2000.

Early Attempts to Deal with Toxic Chemicals

In 1976, six years after Earth Day [April 22, 1970], Congress acted to address the threat from these new chemicals and their introduction into the environment. The Toxic Substances Control Act (TSCA) established methods for identifying chemicals that could pose risks to humans, plants, and animals—and placed controls on their manufacture, distribution, use, and disposal. The Resource Conservation and Recovery Act (RCRA) provided a framework for ensuring the safe disposal of wastes that threaten human health or the environment because they are flammable, explosive, corrosive, or toxic. RCRA required that such "hazardous wastes" be tightly managed from generation to disposal.

TSCA and RCRA addressed the new threats posed by industrial practices developed during the 20th century. Together, they empowered EPA to establish a regulatory scheme to provide protections from the introduction of dangerous chemicals and chemical by-products into the environment.

But Love Canal exposed a gap in this new blanket of protection. Toxic chemicals did not need to be newly introduced to provide a threat to a community. Wastes that had been buried long ago—and mostly forgotten—could suddenly prove to be dangerous.

A new threat to human health and the environment was discovered in the decade after Earth Day. And new ways needed to be developed to address this serious challenge.

Love Canal grabbed the Nation's attention, but it was not alone.

In 1979, EPA estimated that there were thousands of inactive and uncontrolled hazardous waste sites in the United States that could pose a serious risk to public health.

"Wastes that had been buried long ago—and mostly forgotten—could suddenly prove to be dangerous."

Hazardous waste disposal sites were only one part of the problem. Chemical spills posed another danger. Thomas C. Jorling, EPA's top official for waste management, told a Senate committee in 1979:

> Spills of hazardous substances can have serious environmental and public health impacts similar to abandoned hazardous waste disposal sites. Environmental damage resulting from such spills can result in massive fish kills, destruction of wildlife, air pollution, and loss of livestock by contamination of drinking water. Spills have also resulted in loss of life and posed direct threats to human health from toxicity, fires, and explosions.

On April 22, 1980, the Nation celebrated the 10th anniversary of Earth Day. Thousands took to the streets to reaffirm the country's commitment to protecting the environment. But the celebration was tempered by an event that took place the previous evening.

An explosion in a warehouse ignited a fire that burned 24,000 barrels of chemicals, including illegally stored toxic wastes. The fire burned for 10 hours—sending a thick black plume of smoke and ash over a 15-square mile area and raised fears of widespread chemical contamination. The site was completely destroyed and there were reports of burning waste drums launching 200 feet through the air and bursting into cascades of flashing light. Public schools in Elizabeth, Linden, and Staten Island were ordered closed as State authorities urged residents to shut all doors and windows and remain inside. A 72-hour ban on commercial and sport fishing, covering a 40-mile radius, was also imposed.

"By 1980, the decades-old legacy of industrial waste was clearly presenting the Nation with a major problem."

In an April 23 editorial, the *New York Times* commented that the 10th anniversary of Earth Day "got off to a poisonous start" because of the fire in Elizabeth, New Jersey, but that "it, more than any other Earth Day observance, focused attention on the problem of getting rid of toxic wastes." The *Times* further commented that "[t]he dump in Elizabeth is one of those 'ticking time bombs' that environmental officials keep warning us about" and that the accident in New Jersey underscores "the need for long-pending Federal legislation to provide a 'super-fund' for cleaning up hazardous waste sites whose owners can't be found or who shirk responsibility." The *Times* editorial ended by warning: "The Elizabeth site was one of the worst. It is by no means one of a kind."

Congress Creates a "Superfund"

By 1980, the decades-old legacy of industrial waste was clearly presenting the Nation with a major problem. EPA's Thomas C. Jorling declared the Carter Administration's position that, "[r]eleases of hazardous wastes from abandoned and inactive disposal sites are perhaps the most serious environmental problem facing the Nation today." Campaigning for the Presidency, Senator Edward Kennedy (D-Massachusetts) called the disposal of hazardous waste "a public health nightmare of extraordinary dimensions" causing millions of Americans to take "unwitting, involuntary but potentially serious health risks every day, simply because of where they live."

Although the problem was serious, in 1980, the country had few means to address it. Individuals could sue in court for injuries suffered from industrial wastes, but this was costly and time-consuming—and awards were uncertain. More important, any remedy was after-the-fact. The common law did not provide a means to prevent hazardous waste injuries from happening in the first place.

Some of the Federal legislation passed in the wake of the first Earth Day helped to fill this gap—but only partially. RCRA provided EPA with authority to sue owners of inactive hazardous waste sites to prevent "an imminent and substantial danger to human health or the environment." However, this required EPA to identify a person or business in the position to stop a spill from happening. Since many of the sites had been abandoned long ago, such an individual or business often could not be identified. The Clean Water Act established a control program for certain spills of oil and hazardous substances, but this was limited to discharges into navigable waters. The Clean Water Act did not cover spills of hazardous substances onto soils—and only certain designated hazardous substances could be regulated.

The range of problems explored by Congress was addressed by Senator Robert Stafford (R-Vermont) when the Environment and Public Works Committee held its first hearing in 1979 on the possible dangers posed by toxic waste sites:

> If these hearings were to deal only with Love Canal or Toone, Tennessee, we would be neglecting the radium sites in Denver. And if we were to deal with the Denver sites as well, we would still be neglecting PCBs in the Hudson River and PBBs in Michigan. If we restrict ourselves to just waste, we will leave a large gap because in the chemical business one man's meat is literally another man's poison. Waste from one company is feedstock to another. What we must explore is the entirety of how and why toxics are entering the environment, whether they are injuring people, and if so, how. Then we must decide whether there should be a scheme to compensate victims, and if so, for what injuries.

The Senate Environment and Public Works Committee held 11 days of hearings in 1979. In the House, two separate committees held hearings and proposed separate bills for dealing with different aspects of the larger hazardous substances problem. On September 19, 1980, after often-contentious negotiations, the House passed a bill proposing a "superfund" to deal primarily with chemical emergencies.

The Senate meanwhile developed its own "superfund" bill to deal with emergencies, but which also allowed injured parties to sue in Federal court for damages. This bill languished in the Senate until after the 1980 Presidential elections. In November, Senator Stafford introduced an amended proposal. It was a version of this proposal that was eventually enacted.

On December 11, 1980, President Jimmy Carter signed the new Comprehensive Environmental Response, Compensation, and Liability Act of 1980 (CERCLA or Superfund). Calling it "landmark in its scope and in its impact on preserving the environmental quality of our country," President Carter stated that it "fills a major gap in the existing laws of our country."

Comprehensive Coverage

If there was such a thing as a "truth in labeling" requirement for statutes, Superfund would be one law that would meet it. For (as passed by Congress in 1980 and strengthened by amendments in 1986), CERCLA is truly a:

- Comprehensive,
- Environmental Response,
- Compensation, and
- Liability Act.

Congress recognized that the problem was broad—and that broad solutions had to be created. Love Canal showed what could happen with the improper disposal of chemical wastes, but the issue was bigger than that. As stated by a 1980 Senate Environment and Public Works Committee report:

> When confronted with an incident of toxic chemical contamination, it is often difficult to distinguish whether it is the result of a spill, a continuing discharge, an intentional dumping, or a waste disposal site. Any legislative solution would also have to address, in addition to disposal sites, the closely related problems of spills and other releases of dangerous chemicals which can have an equally devastating effect on the environment and human health.

Therefore, CERCLA provides comprehensive authority for the government to act. EPA can respond to:

• A "release" or "substantial threat" of a release of a "hazardous substance" into the environment; or

• A "release" or "substantial threat" of a release of "any pollutant or contaminant which may present an imminent and substantial danger to public health or welfare."

"Release" includes virtually any situation where a hazardous substance is released from its normal container. "Substantial threat of release" is even broader—allowing EPA to respond in situations like corroding tanks or abandoned drums, where there is even a risk of "release."

EPA may respond to an actual or potential release of any quantity of a "hazardous substance" or "pollutant or contaminant" in two general ways:

• Removals; or

• Remedial actions.

Removals deal primarily with environmental emergencies—and are generally short-term actions to diminish the threat of a release. Examples include cleaning up waste spilled from a container, building a fence around a site, or providing fresh water to residents whose regular water supply has been contaminated.

Remedial actions are long-term, permanent cleanups. Examples include excavating waste and transporting it to a facility that can safely handle it, treating the waste to remove contaminants, or placing clay covers over or barriers around the waste to prevent migration. Remedial actions may take many years and cost millions of dollars, in order to make the site safe for human health and the environment.

"Love Canal showed what could happen with the improper disposal of chemical wastes."

Most of the 1980 press coverage about the passage of CERCLA concentrated on the Superfund Trust Fund, which gave the statute its nickname. The Trust Fund is financed from various taxes and court awards from the parties found responsible for hazardous substances releases. The 1980 law authorized a Trust Fund of $1.6 billion. The 1986 amendments to CERCLA increased this amount to $8.5 billion.

The Trust Fund can be used to address both emergencies and longer-term cleanups. It can pay for both actual cleanup costs and for EPA's enforcement actions. It also is available to pay for certain natural resource damages, reimbursement of local governments, and claims by private parties.

> ***"The [Superfund] Trust Fund can be used to address both emergencies and longer-term cleanups."***

Many times, the Trust Fund provides financing so EPA can address a hazardous substance release first, rather than have to wait for a court to determine who was responsible for causing the release. Later, when the court determines who is liable, EPA recovers its response costs and the Trust Fund is reimbursed. This is one of the major innovations of CERCLA since, prior to the statute's enactment, the common law required that liability be determined first before any action could be taken.

EPA has three basic options when it responds to a release:

• Conducting the cleanup itself using money from the Trust Fund and then seeking to recover its costs from the potentially responsible parties (PRPs);

• Compelling the PRPs to perform the cleanup through administrative or judicial proceedings; or

• Entering into settlement agreements with PRPs that require them to clean up the site or pay for cleanup.

In all cases, the responsible party pays since CERCLA provides EPA with strong enforcement authorities. Congress decided that the parties who created these sites should be the ones who pay for cleaning them up.

Congress passed a Superfund statute, but it was up to EPA to create a Superfund program.

Because of national media attention on the problems at Love Canal, the Valley of the Drums, and other high profile sites, immediate and effective action was expected of EPA. Drums had to be collected and removed. Fires extinguished. Leaks from tanks and waste ponds stopped.

But responding to spills was not enough: EPA needed to clean up sites so they would continue to be safe in the future.

In order to make the Superfund program effective for the long-term, a large investment of resources was needed. EPA had to create a regulatory framework to carry out the mandate of Congress. This had to be done even though EPA faced a series of unknowns. The health effects of chemicals needed to be researched. Technologies had to be created to safely treat, store, and dispose of wastes. There was a general lack of data about specific sites—coupled with a fledgling scientific understanding of waste migration. There also was a shortage of trained personnel, such as engineers, to address these problems.

Nothing like Superfund had ever existed before. Over time, a strong and effective program evolved to protect human health and the environment from the dangers of hazardous wastes. . . .

By the close of Superfund's first decade, it became clear that the goal [of a clean and safe environment] could not be achieved simply by laws and regulations—or by the Federal government alone. Instead, partnerships needed to be formed. EPA reached out to States, Tribes, communities, and industry to forge stronger relationships.

EPA facilitated these partnerships through reform of existing programs and creation of new innovative ones. The goals of protecting human health and the environment remained the same, but the means were reinvented. Today, Superfund is more flexible, more effective, more sensible, and more affordable—seeking to achieve the best environmental results for the least cost.

But the proof of Superfund's success is found in our backyards.

The Industri-Plex site in Woburn, Massachusetts is one illustration of what can happen when partnerships are formed among the community, State, EPA, and the private sector. Industri-Plex is a 245-acre industrial park located 12 miles north of Boston along the heavily-traveled Interstate-93 corridor. Since 1853, it had been the home of various chemical manufacturing operations, including the manufacture of glue from raw animal hides and chrome-tanned hide wastes. These operations caused the soils and the ground water to become contaminated with elevated levels of metals, such as arsenic, lead, and chromium.

Industrial activities ceased at the site in 1969, and the property was sold for development. In the late 1970s, the developer unearthed animal hides, which emitted odors that smelled like rotten eggs. Because of community protest, development activities ceased at the site in 1980 and the Federal government became involved. The site was placed on the first National Priorities List (NPL) in 1983. [The NPL identifies sites that are national priorities for receiving further investigations and long-term cleanup actions.]

Because of innovative thinking and flexibility, a site that was once the subject of community unrest has been transformed into a center of community pride. When the Federal government settled with the PRPs in 1989, two Trusts were formed among EPA, the Massachusetts Department of Environmental Protection, the City of Woburn, and 24 current and former landowners. The Trusts facilitated the cleanup of the site and its eventual redevelopment.

"A site that was once the subject of community unrest has been transformed into a center of community pride."

The many partners were committed to making the Industri-Plex site both safe and economically viable. Lines of communication were kept open, and ways to resolve normally difficult problems were found. Today, this former toxic wasteland has been cleaned up and redeveloped for the following uses:

• *Regional Transportation Center*—State agencies have developed a 36-acre transportation facility that can accommodate 2,400 parking spaces for commuter train service into Boston, a Park and Ride bus service, and shuttle service to Logan Airport.

• *Commercial and retail district*—A Target department store has been constructed, along with 750,000 square feet of office and hotel space.

• *New highway links*—A new highway interchange between I-93 and I-95 eases severe regional traffic congestion and provides access to new businesses. Additionally, the main thoroughfare through the site has been improved and extended.

"Superfund will continue its evolution to meet the new challenges of a clean and safe environment."

Cleaning up Industri-Plex has been good for the environment, but it is also a boon to the local economy. The new developments at Industri-Plex now provide as many as 4,300 permanent jobs, approximately $147 million in annual income associated with those jobs, and a $4.6 million potential increase in residential property values within two miles of the site.

Creating a New Wildlife Habitat

Superfund—combined with innovation, communication, and partnerships—can also lead to new environmental habitats.

The 12-acre Bowers Landfill in Circleville, Ohio, was first operated as a pit for gravel excavation, but it was converted to a municipal solid waste landfill. Later the landfill began accepting industrial wastes, including approximately 7,500 tons of chemical waste.

Disposal practices at the Bowers Landfill frequently consisted of depositing waste directly onto the ground and covering it with soil. Waste also was burned on-site. In 1980, investigations determined that contaminants in the landfill were polluting nearby monitoring wells with volatile organic compounds. In 1983, the site was added to the first NPL.

Partnerships formed quickly once the site was identified as a national priority. The partners included EPA, the Ohio Environmental Protection Agency, the Ohio Division of Wildlife, the U.S. Fish and Wildlife Service, and the local community. Each partner played a crucial role in the planning and design of the cleanup. In 1985, the Bowers Landfill Information Committee was formed, providing the surrounding community with an opportunity to become involved with the daily activities of the site. These partnerships facilitated communication, which in turn fostered numerous positive economic and social impacts for the local community.

Cleanup of the Bowers Landfill required many creative innovations. For example, EPA and the State of Ohio decided that they needed to do something to protect the newly capped landfill from floodwaters that frequently inundate the land along the Scioto River. The site's location near the river made it ideal for creating wetlands.

This innovative and cost-effective use of the land not only controls flooding, but benefits the surrounding ecosystem. The wetlands are now flourishing, providing a safe habitat for numerous species of plants, birds, and other wildlife.

Wildlife habitats. Transportation centers and shopping malls. These are just some of the successes of Superfund.

As Superfund enters its third decade, EPA faces four central challenges:

• The Agency will continue the cleanup of NPL sites, as well as address immediate contamination problems through Removal Actions across the country;

• EPA will continue to ensure that cleanup remedies remain protective of human health and the environment for years to come;

• As new sites are identified, EPA will share responsibility with States, Tribes, and other stakeholders to work with communities and PRPs to get these sites cleaned up efficiently; and

• The Agency will continue to serve as a catalyst to promote redevelopment at both brownfields and former Superfund sites.

Because of Superfund, sites that were once dangerous have been made safe. Land that was once desolate has been restored to productive use for communities across America. And new toxic waste sites are prevented from occurring in the first place by the presence of Superfund.

This is Superfund on its 20th anniversary. Now entering its third decade, Superfund will continue its evolution to meet the new challenges of a clean and safe environment—the promise of Earth Day.

Mandatory Pollution Regulations Are Effective

by Peter K. Krahn

About the author: *Peter K. Krahn is the acting head of the inspections division of Environment Canada for the Pacific and Yukon region. Environment Canada is the government agency responsible for enforcing Canada's environmental laws.*

Governments around the world are considering substitutes for monitoring and enforcing environmental law within industries. Many believe the industries that harm the environment should be allowed to monitor themselves and that they will voluntarily comply with the law. Parts of the private sector favour this approach, while many citizens, members of the scientific community and environmental organizations doubt the effectiveness of voluntary compliance, seeing it as no more than wishful thinking on the part of governments. This study provides a factual basis for policymakers when determining the best course of action. Other jurisdictions may find the results of this work useful when analyzing their local or national concerns about adherence to environmental law.

The Case for Mandatory Regulations

Wood preservation facilities and pulp and paper mills in British Columbia (BC) discharged over 750 million cubic meters of toxic effluents in 1986. Concentrations of chlorophenols, copper, chromium, arsenic, creosotes and natural wood extractives in these discharges could cause 100% mortality of salmon in under eight minutes. Pentachlorophenol, a wood preservative that was the most widely used wood fungicide in BC, inhibits the ability of an organism to convert food into energy, resulting in suffocation and heart failure at 30 parts per billion (ppb). Pentachlorophenol was detected throughout the lower Fraser River and in marine waters near sawmills and pulp and paper mills in 1986. Heavy metals such as copper, chromium and arsenic (used in long-term wood preservation) are acutely toxic to fish at concentrations as low as 20 ppb and were discharged

Excerpted from "The Business of Environmental Law," by Peter K. Krahn, *Global Biodiversity*, Fall 1998.

at levels exceeding 9,000 ppb. Dioxins and furans which are bio-accumulative and cause immune system damage, liver dysfunction, impaired reproduction, birth defects and cancer in mammals in the parts-per quadrillion range were detected in all pulp mills using a chlorine bleaching process.

In response to these findings, Environment Canada—responsible for enforcing the Canadian Environmental Protection Act (CEPA) and the pollution prevention provisions of the federal Fisheries Act—initiated the Fraser River Action Plan (FRAP) in 1992. The department's Pacific and Yukon region set out to stop and reverse existing environmental contamination and degradation of the river by implementing strategies to reduce pollution and virtually eliminate the discharge of persistent toxic substances. Four years later, the enforcement division conducted a review of historical data related to the compliance with technical criteria and effluent discharge limits by 19 industrial sectors. This study demonstrated that in most cases voluntary measures achieved a minimal positive result for both compliance rates and reductions in toxic discharges. It showed as well that mandatory environmental law enforcement strategies dramatically improved average compliance rates from a pre-enforcement average of 60% to over 90% and subsequently reduced environmentally harmful effluent discharges by up to 99%.

How a Compliance and Enforcement Program Rolls Out

The examination of 19 industrial sectors in the compliance study revealed distinct phases in an enforcement cycle that will last from five to ten years, depending on the intensity of the program. Compliance history and some random selection determined which facilities were chosen for examination. Phases 1 and 2 involve the scientific assessment of an environmental issue and the development of best practices. Field inspection techniques are designed and tested in phase 3 and implemented in phase 4, where up to 80% of the facilities may be brought to a reasonable level of compliance.

Phase 5 includes strategic enforcement initiatives that may lead to investigation of up to 5% of the facilities in any industrial sector. Prosecution and conviction of up to 2% of the facilities occur in the sixth and seventh phases. Conviction can mean fines of up to $1 million per day for each day of the offence and/or three years imprisonment. Court orders may prohibit businesses from undertaking certain activities or may require them to improve fish habitat, publish facts about the offence, pay compensation, perform community service or post performance bonds. The eighth and final phase involves re-inspections to verify compliance with official warnings, directions and court orders.

Enforcing Compliance in the Wood Preservation Industry

British Columbia supplies an estimated 39% of the world's soft wood lumber supply and annual sales have often exceeded $4 billion, providing major employment and tax revenues. Prior to 1983, water-based solutions of pentachlo-

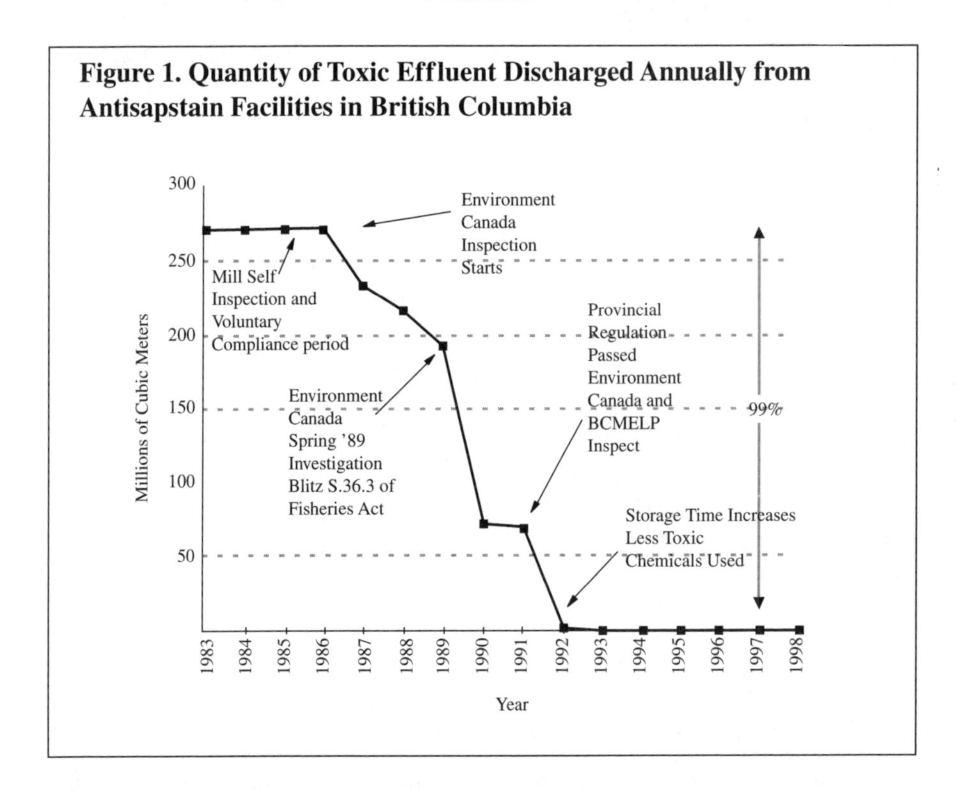

Figure 1. Quantity of Toxic Effluent Discharged Annually from Antisapstain Facilities in British Columbia

rophenol (PCP) and tetrachlorophenol (TTCP), called antisapstain chemicals, were used to treat wood at sawmills to control molds and fungi that attack freshly cut wood. The treated lumber was then moved to exterior storage yards where it was exposed to precipitation. Rainwater leached the chemical from the wood and an estimated 250 million cubic meters of acutely lethal effluent were discharged annually from these facilities into fresh water and marine environments that supported valuable salmon and other fish and shellfish stocks. The tidal effect on the lower Fraser estuary caused storm water discharges to stagnate near outfalls, which created immediate zones that were acutely toxic to fish, and sub-lethal zones of up to 600 meters in length.

"Mandatory environmental law enforcement strategies dramatically improved [industry's] compliance rates [with pollution regulations]."

From 1986 to 1989, Environment Canada initiated a voluntary compliance program using specific checklists and on-site visits combined with compliance promotion seminars (Figure 1). The progressive mills implemented proper chemical handling and treatment procedures or constructed facilities to control or prevent releases. However, a significant proportion of the mills did not implement corrective measures. During this period, charges were not laid for improper practices.

In early 1989, Environment Canada's enforcement staff embarked on a strategic enforcement initiative, targeting five mills for investigation and ultimate prosecution as a result of their severe non-compliance. As part of this enforcement initiative, a federal/provincial committee drafted a regulation, enacted by the BC government in 1991, to make a number of environmentally sound operating practices mandatory. A comprehensive inspection and sampling program by federal and provincial inspectors followed. Under the new regime, incentives encouraged businesses to develop new, less toxic chemicals; and to improve facilities or find alternate markets that did not require treatment. By 1993, the quantity of acutely toxic effluent discharge decreased by an estimated 99%.

BC Forces Pulp and Paper Mills to Comply with Dioxin and Furan Controls

There are 75 variations of dioxins and 135 varieties of a related group of chemicals called furans. The most environmentally disruptive are 2,3,7,8-tetrachloro-paradibenzodioxin (2,3,7,8-TCDD) and 2,3,7,8-tetrachlorodibenzofuran (2,3,7,8-TCDF) which are defined as toxic chemicals under the CEPA. The two primary sources of these chemicals were saw mills and pulp and paper mills that used chlorine in their bleach plants. In the late 1980s, Environment Canada and the Department of Fisheries and Oceans collected numerous sediment, crab and shellfish samples in areas near sawmills and pulp mills. Following the analysis of these samples, 1,200 square km of crab and shellfish harvesting areas were closed.

"From 1986 to 1989, Environment Canada initiated a voluntary [pollution] compliance program. . . . [Most] mills did not implement corrective measures."

New federal pulp and paper regulations passed in 1989 required an immediate ban on the purchase and use of wood products contaminated with chlorophenols as feed stock and prohibited the production, import or sale of defoamers contaminated with dioxin and furan precursors. The mills implemented these bans in anticipation of the mandatory regulations, resulting in a 99% reduction in the discharges of the two regulated chemicals, 2,3,7,8-TCDD and 2,3,7,8-TCDF.

Lethal Discharges from Mills Are Drastically Reduced

Water-based mixtures of copper/chromium/arsenic or oil-based mixtures of pentachlorophenol or creosote are used in the production of pressure—or thermal—treated lumber and poles used for telephone and power distribution lines, railway ties and structural and landscaping materials. The treatment and storage of finished products resulted in annual discharges of up to 1 million cubic meters of contaminated storm wafer runoff from all 18 heavy-duty wood preservation mills in BC. An estimated 600,000 cubic meters per year entered surface and ground water systems of the Fraser River from six mills located near Vancouver.

Environment Canada developed five codes of practice in 1984 that were voluntary measures to be implemented by the industry to prevent spills, control contaminated storm water runoff and contaminated waste disposal. In 1987, Environment Canada informed the industry of the results of studies concerning contamination of soils and, in particular, storm water runoff.

> ***"The period of voluntary compliance resulted in . . . unsatisfactory changes in the quantity of pollutants discharged [by industry]."***

From 1983 to 1991, the industry operated under a voluntary program to implement Environment Canada's code of practice recommendations. In 1991, the department conducted further scientific research that confirmed these mills were still releasing significant quantities of acutely toxic effluent. Each mill was informed of the results, however no significant operational changes occurred under the voluntary program. In February 1994, Environment Canada initiated an intensive inspection and investigation program that targeted all six Greater Vancouver mills for violations of the Fisheries Act. Since 1994, the discharge of acutely lethal effluent from this industry sector has decreased by over 90%. Recent follow-up inspections have verified that there has been a 34–85% reduction in the quantities of environmentally harmful substances in Fraser River sediments adjacent to the five mills where investigations were initiated. . . .

Analyzing the Trends of Industry to Comply

The period of voluntary compliance resulted in negligible or unsatisfactory changes in the quantity of pollutants discharged in the antisapstain industry, with outcomes improving only when an enforcement program enhanced voluntary compliance. Similar outcomes, as in the decline of discharges from the pulp acid paper and heavy-duty wood preservation sectors, resulted when enforcement resources were diverted to deal with these industries.

These observations support the data reported in the 1996 Canadian Environmental Management Survey, which found that the most influential factors for promoting environmental change were: compliance with regulations at more than 90%, board of director liability at over 70%, employees at more than 60%. The least influential factors were: voluntary programs 15–20%, interest groups 10–12% and trade considerations at under 10%.

The Economics of Compliance Promotion and Enforcement Programs

There are three primary economic issues encountered in the development of a compliance and enforcement program: a) costs to the regulatee (company) to implement corrective measures, b) competition within an industry and c) investments by regulatory agencies in compliance promotion and enforcement programs. Experience indicates that after a strategic enforcement program is initiated, it takes from five to seven years for the industry to make the technical and structural changes necessary to achieve compliance.

"Many people believe that good environmental management practices can be implemented on a voluntary basis by industry. . . . [S]uch a belief is often wishful thinking."

Experience has also indicated that initial estimates of costs to industry to achieve compliance are frequently higher than the actual costs ultimately paid out. For instance, the antisapstain wood preservation industry estimated that environmental improvements necessary to comply with the Fisheries Act would cost an average of $10 million per mill. In 1988 this was revised to $5.3 million in an industry-generated report. The actual costs experienced by 1996 ranged from $1 million to $1.5 million per mill.

Facilities that implement corrective changes can suffer a significant economic penalty if they compete in regional, national or international markets that reflect great disparities in environmental legislation and enforcement. For instance, the average British Columbia pulp and paper mill with a production of 1,000 tonnes per day incurs a daily cost of $30,000 ($9 million annually) to meet environmental standards. If a mill of similar size in another region does not comply with the same standards, it will gain an economic advantage of $9 million or more, depending upon interest rates and capital cost factors.

Implementation of targeted enforcement programs incurs an expense to the government that is distributed to the entire population through taxation. The federal government spent $600,000 in operational costs for the heavy-duty wood preservation enforcement program. The industry costs to comply with the federal and provincial requirements will total $39 million by September 1998. This is nearly a 1:70 ratio of federal government expenditure provoking industry expenditure. Under targeted enforcement programs, the costs of environmental improvements are initially borne by the facility and then passed on to the consumer of the particular product. In this manner, the "polluter pay" principle is directed more to the producer and consumer of the products; whereas should the government offer a grant or tax benefit program, that would distribute the costs to the public at large, many of whom may not use the products.

In order to maintain and sustain a healthy ecosystem, we must reduce the quantity of toxic and environmentally disruptive chemicals that are discharged, thereby enhancing habitat for people, plants and animals. Many people believe that good environmental management practices can be implemented on a voluntary basis by industry. But as the evidence provided in this paper demonstrates, such a belief is often wishful thinking and is rarely borne out in the real world.

The compliance rates of industrial sectors engaged in voluntary compliance programs in British Columbia averaged a 60% implementation rate of best management practices, frequently resulting in the discharge of toxic or environmentally harmful effluents. Regulated industries that were subjected to federal (or combined federal/provincial) inspections and/or four to five years of sustained enforcement initiatives averaged a 94% compliance rate. Furthermore, among these industries, the discharges of harmful substances frequently decreased by over 90% from the pre-enforcement period. These trends were observed in 19 industrial sectors, which indicate that a strong enforcement program is necessary to provide incentives for voluntary initiatives.

Good managers will pursue methods that are beneficial on both a cost and environmental basis, but they will encounter barriers to this dual effort that are imposed by economic competition. The nature of the competitive environment makes industry owners and/or managers poorly suited to impose environmentally responsible operating conditions on recalcitrant members of their particular sector. However, properly designed environmental law enforcement programs level the economic playing field, thus reducing the impact of competition in society's efforts to protect the environment by focusing the cost of pollution control on the producers and consumers of polluting products and services.

Clean Water Regulations Should Be Strengthened

by Stephen B. Lovejoy and Jeffrey Hyde

About the authors: *Stephen B. Lovejoy is a professor of agricultural and environmental policy at Purdue University. At the time this viewpoint was written, Jeffrey Hyde was a graduate research assistant in the department of agricultural economics at Purdue University.*

Essentially all surface water in the United States is polluted, much of it from industrial waste and the discharge of municipal sewage. Over the past few decades, such sources of pollution, known as point sources, have been substantially reduced as a result of stricter regulations. The nation's waters still have not met our quality goals, however, mainly because other sources of pollution—known as nonpoint sources—have yet to be curbed.

"Nonpoint" Sources of Water Pollution

Stormwater runoff is the dominant means by which nonpoint-source pollutants reach surface water. Such runoff carries pollutants from several land-use activities.

For example, silviculture, or cultivation of timber, is responsible for water pollution in nearly 40 states. Each step of the timbering process—from planting to cutting—allows pollutants to escape into surface waters. The major pollutants from forests are sediment, fertilizers, and pesticides.

Mines are another major source of sediments. The by-products of mining—acids, heavy metals, salts, and radioactive materials—can seep into groundwater before flowing into surface waters. Under the Clean Water Act of 1972, only abandoned mines are regarded as nonpoint sources of pollution from mining. Indeed, operating mines are considered to be point sources and therefore are not regulated by nonpoint-source programs.

Meanwhile, the construction of buildings, roads, houses, and commercial developments disturb soil and allow more than the normal amount of sediment to

From "Nonpoint-Source Pollution Defies U.S. Water Policy," by Stephen B. Lovejoy and Jeffrey Hyde, *Forum for Applied Research and Public Policy*, Winter 1997. Reprinted with permission.

be washed away with rainfall. Stormwater runoff from urban areas also carries oil, litter, and chemicals into surface water, either directly or through urban stormwater collection systems.

While all these activities contribute significantly to nonpoint-source pollution, the main source of such pollution is agricultural land use. In fact, the U.S. Environmental Protection Agency (EPA) contends that agricultural land is responsible for polluting or restricting the use of more than 70 percent of U.S. rivers and 50 percent of U.S. lakes. Agriculture is a source of many different pollutants, including pesticides, fertilizers, animal wastes, and sediments.

Our knowledge of the contribution these sources make to water pollution has increased as point sources of pollution have been reduced. And as more and more of our nation's waste pipes have been capped or filtered, legislation has become increasingly aimed at controlling nonpoint-source pollution.

Tracking Nonpoint-Source Pollution

Nonpoint-source pollution differs from point-source pollution in three ways. First, monitoring individual contributions from each nonpoint source is difficult, if not impossible. For example, a researcher cannot say with certainty how much phosphorus ran off Ms. Smith's cropland before spilling into the river.

The second feature is that the causal link between a farmer's level of chemical inputs, management practices, and the resulting load of water-borne pollutants is unclear. It is difficult, for instance, to attribute an increased amount of pollution runoff to one crop instead of another, to a new combine or other piece of equipment, or to an increase in the use of agricultural chemicals.

"The nation's waters still have not met our quality goals."

And third, such random variables as wind, rain, and temperature also affect the load of nonpoint-source pollution making its way into surface water. For instance, temperature and rainfall affect the breakdown of agricultural chemicals and the ability of the soil to hold moisture.

While the sources of nonpoint-source pollution are known, less is known about the impact these pollutants have on the nation's water resources. For instance, over time these pollutants can adversely affect the respiration and reproductive patterns of fish and other aquatic animals. Furthermore, some pollutants can accumulate in fish, leading to health risks for the humans who consume them.

Many pollutants also may pose health risks for humans if they are found in drinking water at higher concentrations. Beyond health concerns, nonpoint-source pollution restricts such recreational activities as boating and fishing and detracts from the beauty of our waterways. And as recreation, drinking-water quality, and aesthetic qualities are harmed, land values decline.

In 1962, [biologist] Rachel Carson's *Silent Spring* brought the distressing impacts of agricultural pollution to the attention of the nation. And by the late

1970s, studies had shown that erosion and phosphorus from animal waste were major factors in eutrophication—the process of lake aging—by reducing the levels of dissolved oxygen in the water. The studies also demonstrated that agriculture was responsible for high levels of sediments, phosphorus, and nitrogen in surface water.

"[The Clean Water Act] helped stabilize the amount of pollution entering the nation's waterways."

A decade after publication of Carson's book, Congress amended the Federal Water Pollution Control Act of 1948. The amendments, which quickly became known as the Clean Water Act, marked the beginning of major legislation aimed at controlling and eliminating water pollution. The goal was to provide Americans with clean water for fishing and swimming.

This legislation was more focused on controlling point sources than on nonpoint sources of pollution because the former were perceived as a larger threat to the nation's water and would be easier to control. The act proved largely successful in controlling point sources of pollution; overall, it helped stabilize the amount of pollution entering the nation's waterways.

While earlier amendments targeted point sources of pollution, the 1977 amendments were aimed more at nonpoint sources. Specifically, they required each state to design and implement plans to control all pollution sources, though the main emphasis fell on nonpoint sources. Furthermore, the governor of each state was to designate at-risk bodies of water as prime targets for the implementation plans. The plans were to include three main elements.

First, they would identify projects, such as secondary-treatment facilities or cooling ponds, that were deemed necessary to meet municipal and industrial waste-treatment goals. Second, each plan would identify key nonpoint sources of water pollution. Third, the plans would include specific procedures for controlling these sources. Responsibility for regulating these programs was usually placed in the hands of local governments, often at the county level.

The 1977 amendments also gave authority to the U.S. Department of Agriculture's (USDA) Soil Conservation Service, now known as the Natural Resources Conservation Service, to design a program for so-called best management practices. Farmers themselves were to take a direct hand in implementing these efforts.

Best management practices might include, for example, crop rotation, which helps to maintain natural nutrient levels and thereby decreases the need for fertilizers. In other cases, these practices might involve conservation tillage, which helps minimize ploughing and thus reduces soil erosion. In still other cases, the practices might rely on integrated pest management, which seeks nontoxic measures to control pests and thus decrease pesticide use.

The Clean Water Act was reauthorized and amended again in 1987. An important provision of these latest amendments required states to identify rivers and lakes whose environmental quality would likely decline without implementation of nonpoint-source pollution controls. Despite their ambitious intent, however, the 1987 amendments have fallen short of their goals. Indeed, agricultural nonpoint-source water pollution remains a major problem in many areas of the country. In fact, this form of pollution continues to restrict use of more than half of the lakes and miles of river in the United States.

Though the 1987 amendments failed to achieve complete success, they established the State Revolving Loan Fund. Congress has appropriated more than $10 billion to the fund, which requires states to pay 40 percent of the costs of constructing, for example, secondary sewage-treatment facilities and of improving combined stormwater systems.

Though the current appropriations are by some standards generous, they are only about half what the states claim they need to meet their goals in terms of controlling nonpoint-source pollution.

Some observers insist that the Clean Water Act has failed to accomplish its goals, in part because EPA has not been given adequate enforcement tools to ensure that the goals are being met. Specifically, critics cite the inability of government agencies to bring civil or criminal actions against nonpoint polluters. The only enticements, in fact, have been incentive-based. The bottom line is that the Clean Water Act has met with only limited success in reducing and eliminating nonpoint sources of pollution.

Safe Drinking Water

The Safe Drinking Water Act of 1974, established with the goal of providing the nation with contaminant-free drinking water, instituted standards that have become ever stricter over time. Although the act has achieved some success, EPA has often failed to administer the law in a consistent and effective manner.

The act's first phase targeted waters that had the highest contaminant levels from point and nonpoint sources and that posed a direct threat to human health. Phase two, which is more regulatory in nature, prescribed maximum-allowable contaminant levels and treatment methods for all community water systems.

Phase three targeted aesthetic and other nonhealth-related pollution issues. Specifically, its goal was to remove odors and coloring that may cause some users to substitute bottled water for tap water even though the tap water poses no added health risks.

One of the unique features of the Safe Drinking Water Act is that EPA's administrator has the authority to bring civil action against individuals or institutions that fail to comply with the regulations or are not meeting the regulations according to schedule.

When the act was reauthorized by Congress in 1996, it included the following two reforms. First, EPA was relieved of its commitment to regulate 25 new contaminants every three years. Instead of being held to a fixed number, EPA has the flexibility to consider any number of contaminants and decide whether they require additional study or regulation.

Second, the act established a revolving loan fund similar to the one implemented under the Clean Water Act. The fund amounted to $1.3 billion in fiscal year 1997. The act also called for creation of a program under which states would be given authority to collect user fees based on the quantity of water that an industry uses or the amount of pollutants it releases.

Water Quality Initiative

The U.S. Department of Agriculture's (USDA) 1989 Water Quality Initiative represents the federal government's response to the declining quality of the nation's surface and groundwaters. In particular, it represents a "coordinated effort to protect the nation's water from contamination by agricultural chemicals." The goal of the initiative is to give agricultural workers the know-how, technology, and funding to prevent pollution problems.

Under the initiative, several programs were implemented to help reduce agricultural nonpoint-source pollution:

• *Assistance.* Under this program, the USDA provides farmers with financial and technical assistance to meet water-quality goals that the state sets forth. These technologies include reduced tillage, integrated pest management, and nutrient and manure testing.

Thirty-seven areas were originally selected for inclusion in the program, based on three criteria: (1) the role agriculture plays in causing pollution; (2) the amount of agricultural pollution that comes from a group of designated pollutants, including pesticides, nutrients, and animal wastes; and (3) the extent to which the area has conformed with other water-quality programs.

While another 275 projects were to be added to the program, budget constraints have prevented the USDA from reaching that goal.

• *Demonstrations.* Under this program, 24 projects were chosen to demonstrate the efficacy of selected, cost-effective practices designed to reduce agricultural nonpoint-source pollution. These practices included curbing the use of nitrogen pesticides through integrated pest management and constructing new livestock watering sites.

Few data are currently available to measure the success of this program. The initiative is just a few years old, and it is designed to remedy a problem that took decades to unfold.

"Agricultural nonpoint-source water pollution remains a major problem."

Nevertheless, preliminary research on 16 projects suggests that nitrogen use has been cut by 7 million pounds and phosphorus use by nearly 4 million pounds. While these are not large reductions compared to the amount that farmers use, they appear to be a step in the right direction.

The 1996 Farm Bill, also known as the Federal Agricultural Improvement and Reform Act, authorizes more than $2.2 billion in additional funding for conservation programs, including programs to reduce nonpoint-source pollution.

A key provision of the bill, the Environmental Quality Incentives Program (EQUIP), provides technical, financial, and educational assistance to farmers who institute certain conservation practices that improve soil and water quality. Program funds, which will total $200 million annually until 2002, are divided equally between crop and livestock operations.

The program targets those agricultural areas that pose the greatest risk to soil and water quality. High priority areas will receive 75 percent of EQUIP dollars, while the remaining funds will be used to meet environmental objectives in other regions.

Clarifying the Issue

A common cord runs throughout our nation's efforts to curb nonpoint-source water pollution: the programs have been, at best, marginally successful in dealing with pollution. The major reason in each case is the incongruence of costs and benefits. Indeed, it's often the case that those who benefit from efforts to clean up our nation's water are not the ones who bear the costs.

For instance, farmers must pay the costs of altering their inputs or management practices, even though they may actually lose money by implementing these changes. Indeed, those who benefit the most are the fishermen, swimmers, boaters, and others who are presently not using the water because of nonpoint-source pollution.

> ***"The future of the nation's rivers and lakes will depend on policies that reduce the quantity of pollutants that . . . enter surface and groundwaters."***

In the future, all programs aimed at improving our nation's water quality should take into account the incentives facing decision makers as well as the issue of who benefits and who pays. In some cases it might make sense to link available funds to a program's success. In other cases, it might be necessary to threaten those who fail to comply with civil or criminal action. Either way, the future of the nation's rivers and lakes will depend on policies that reduce the quantity of pollutants that currently enter surface and groundwaters.

Few resources are as vital to life as water. If we fail to ensure its purity, we will play a hand in poisoning not only our own species but all creatures that inhabit our Earth.

Clean Air Regulations Protect Public Health

by Carol Browner

About the author: *Carol Browner served as administrator of the Environmental Protection Agency from 1993 to 2001.*

Editor's note: The Environmental Protection Agency (EPA) issued new national air quality standards in 1997, setting tighter limits on ground-level ozone (smog) and fine particles (soot) under the Clean Air Act. Many industries, including chemical and trucking companies, opposed the new regulations and sued the EPA on the grounds that they would be too costly to implement. In February 2001, the U.S. Supreme Court unanimously ruled in favor of the EPA, affirming the EPA's authority to set national air quality standards solely on the basis of health considerations. As of January 2002, it remained unclear as to whether the Bush administration would direct the EPA to enforce the new regulations.

On July 19, 1997, the Environmental Protection Agency (EPA) issued its updated air quality standards for ozone and particulate matter—otherwise known as smog and soot. These standards are the culmination of the most thorough scientific review process in EPA history. Mountains of evidence—all of it thoroughly peer-reviewed by scientists, industry experts, and public health officials—led us to the conclusion that air quality standards developed in the 1970s had to be updated for the 1990s because they were not protective enough and too many Americans faced health risks under them. The final product—the first revision in the ozone standard in 20 years and the first-ever standard for fine particulates—is a major step forward for protecting the public health in this country.

The Evidence

An overwhelming body of independently reviewed evidence tells us that the existing standards for smog and soot are not sufficient to protect the public's health with an adequate margin of safety.

From "Smog and Soot: Updating Air Quality Standards," by Carol Browner, *Public Health Reports*, September/October 1997.

For example, the evidence shows that repeated exposure to ozone pollution at previously acceptable levels can cause permanent damage to the lungs and that children, exercisers, and outdoor workers face particular risks. Studies have linked ozone to aggravated asthma in children and adults, to increased emergency room visits and hospital admissions, to reduced immune defenses, and to temporary reductions in lung capacity of 15% to 20% in healthy adults. One study linked ozone with lung damage equal to more than half that experienced by pack-a-day smokers, while other studies found consistent loss of lung function in children playing outdoors in the summertime. Emphasizing the importance of these findings, the American Academy of Pediatrics has recommended that pediatricians should advise parents to keep their kids indoors when ozone levels are high.

"[Updated air quality standards are] a major step forward for protecting the public health in this country."

Also compelling is the scientific evidence on fine particulates, which penetrate deep into the lungs. Each year, thousands of Americans, most of them elderly, die prematurely from respiratory illnesses and heart attacks linked with exposure to them. The American Cancer Society found that the risk of early death is 15% to 17% higher in areas where levels of fine particulates are the highest, while another study showed that individual lives are shortened an average of one to two years in the most polluted cities.

Lung disease is the third leading cause of death in this country—killing an estimated 335,000 Americans each year. Asthma is the most common chronic illness in children, who breathe in nearly 50% more air per pound of body weight than adults. Asthma is now the leading cause of hospital admissions for children, and deaths from asthma attacks among children and young people more than doubled between 1980 and 1993.

Setting the Standards

Are we ready to abandon the nation's long-standing commitment to a health-based standard for air pollution? The public debate over these new standards focuses on a simple question: are we prepared to trade the health—even the lives—of large numbers of people because a few industry spokespeople project "high costs" to reduce their pollution of the public's air?

I believe the answer is no. Americans want clean air. They want their children protected. They want EPA to do its job—ensuring that the air they breathe is safe and healthy. They want EPA to never let up in carrying out its responsibility to ensure that the nation's air quality standards reflect the best and latest scientific evidence about the health hazards of major pollutants.

That is why Congress built into the Clean Air Act a process designed to ensure that air quality standards would be set and, if necessary, revised in a manner that puts the public health first and ensures that Americans are protected with an adequate margin of safety.

Congress wanted to be sure we would never get to the point where the government tells Americans that their air is healthy to breathe while the scientific community knows that, in fact, it is not. Thus the law directs EPA to review the public health standards for the six major air pollutants at least every five years in order to ensure that they reflect the best current science. It also lays out specific procedures to obtain the latest scientific findings and, if needed, to revise the standards.

The process next requires that EPA's standard-setting work and the underlying health studies be independently reviewed by a panel of scientists and technical experts from academia, research institutes, public health organizations, and industry. Once standards are proposed by EPA, they are subject for a period to public comment, after which comments are analyzed and final standards are set. In the most recent standard-setting, the standards were subject to an extra level of thorough review in a Federal interagency process designed to elicit concerns from other parts of government. Congress envisioned that this extensive and comprehensive process would protect the public from the health hazards of breathing polluted air.

A spirit of bipartisanship launched the Clean Air Act in 1970 with an uncompromising promise of public health protection. Then-president Richard Nixon proclaimed it to be "a historic piece of legislation that puts us far down the road toward . . . [the] goal of clean air." The same bipartisan spirit led to the strengthening of the Act in 1990, with President Bush saying that "every American expects and deserves to breathe clean air." And, in fact, due to the success of the original Act and subsequent revisions, many millions of Americans today are breathing healthier air. Millions more of our children are protected from the harmful effects of air pollution.

"The Clean Air Act has . . . protected the public health without holding our economy back."

The Clean Air Act has worked for America. It has protected the public health without holding our economy back. In fact, since 1970, emissions of the six major air pollutants have dropped by 29% while the population has grown by 28% and the gross domestic product has nearly doubled. Time and time again, American industry and the American people have risen to the challenge of cleaner air.

Public Health Comes Before Costs

The Clean Air Act does not allow EPA to consider costs at the critical public health stage of the standard-setting process, requiring instead that pollution limits be based solely on health, risk, exposure, and damage to the environment, as determined by the best available science.

This is no accident. In the 1970 Clean Air Act debate, Congress deliberated the issue of cost in addition to the technical feasibility of meeting clean air standards. The decision was made that the public health must come first. The current best science must prevail in determining the level of protection the public will be guaranteed. Nothing else can take precedence.

> ***"The current best science must prevail in determining the level of [clean air] protection the public will be guaranteed."***

This issue has been revisited both times the Clean Air Act has been amended—in 1977 and 1990. And, each time, Congress and the President have come down firmly on the side of putting the public health first. Not only does the law forbid us from considering costs in setting these standards, but history and real experience tell us we'd be foolish to try.

Almost every time we have begun the process to set or revise air standards, the costs of doing so have been grossly overstated—by both industry and EPA. Dire predictions of economic chaos—always a part of the clean air debate—have never come to pass. Why? Because industry ultimately rises to the challenge, finding cheaper, more innovative ways of meeting the standards and lowering pollution.

At the implementation stage, it is certainly appropriate, under the law, to consider the costs of compliance. In fact, EPA has assembled an implementation package for the new air quality standards designed to give states, local governments, and businesses the flexibility they'll need to meet protective public health standards in a reasonable, common sense, and cost-effective way. We will work with all who are affected—state governments, local governments, communities, and businesses large and small—to find the best strategies for reducing pollution, providing the public health protections, and, at the same time, doing everything we can to prevent adverse economic impact.

Should the nation abandon its commitment to a public health standard for air pollution? I think not. I believe the American people want us to work together with public health professionals, state and local governments, and industry to improve air quality so that future generations can breathe a little easier.

The Superfund Hazardous Waste Program Is Ineffective

by Wayne T. Brough

About the author: *Wayne T. Brough is chief economist at Citizens for a Sound Economy, an organization working for less government and lower taxes.*

When [former] President Bill Clinton was first elected, both the new administration and Congress were calling for significant reforms of Superfund, a program created to clean up the nation's toxic waste dumps. "[W]e all know it doesn't work—the Superfund has been a disaster," noted President Clinton. Three Congresses later, however, Superfund remains virtually untouched by reform, despite broad consensus that the program is broken and ill-suited for accomplishing its original mission.

At the most fundamental level, Superfund is a flawed program built from a premise that guarantees its failure. Although created with the best of intentions—to identify and clean up hazardous waste sites—the Superfund program was structured around a faulty mechanism that relies on a self-perpetuating "blame game" to raise the funds necessary to clean up waste sites. A bizarre set of liability rules at the heart of the program has diverted resources away from actual cleanups in favor of costly legal proceedings and enforcement actions that do little to clean up communities. To date, more than 100,000 parties have been dragged into the Superfund net, which has been cast so wide that it includes the Catholic Church, hospitals, schools, and a Girl Scout troop.

A Brief History

In 1978, Love Canal, an abandoned waste site near Niagara Falls, brought toxic waste into the media limelight. Concerns over chemicals seeping into the homes of local residents and the subsequent evacuation of almost 1,000 families garnered headlines across the nation and spurred Congress into action. The

Excerpted from "Issue Analysis 70—Superfund: The Good, the Bad, and the Broken," by Wayne T. Brough, www.cse.org, February 27, 1998.

result was the Superfund program, which was created in 1980 when Congress passed the Comprehensive Environmental Response, Compensation, and Liability Act (CERCLA). The new federal program enacted an extensive liability system and created a public trust fund—the Superfund—to fund cleanups where responsible parties could not be found.

The Environmental Protection Agency (EPA) and other defenders of the program view the liability system as the core of the Superfund program, claiming that it embodies the virtues of the "polluter pays" principle. Through a web of retroactive, strict, and joint and several liability, the Superfund program searches out "potentially responsible parties" (PRPs) who can then be held liable for the costs of a cleanup. This allows the EPA to cast a wide net when looking for parties to pay for a cleanup. Unfortunately, relying on the legal system for revenue is an unfair and inefficient way to raise money. Resources are focused on litigation, not cleanup, with tremendous amounts of money diverted to legal fees, negotiations, and administrative costs. Studies by the RAND Corporation found that up to 36 cents out of every dollar go toward such costs, and that for firms with annual revenues of more than $100 million, up to 60 cents out of every dollar go towards transaction costs rather that hazardous waste cleanup.

In the case of "orphan sites," where no responsible parties are identified, the cleanup is funded through the Superfund trust. The trust fund is financed primarily through excise taxes on petroleum and chemical feedstocks as well as a corporate income tax. Since its inception in 1980, the Superfund has grown from $1.6 billion to more than $15 billion, of which $10 billion had been spent by the end of 1994. Private sector expenditures on Superfund clean-ups added in, a total of more than $30 billion has been spent over the last 17 years.

Unfortunately, there is little to show for all the money that has been spent. The cleanup process is slow and tedious. On average, the typical Superfund site takes more than 10 years to cleanup at a price tag of $32 million—excluding litigation and administrative costs. As a result, only around 40 percent of the priority sites identified by the EPA have been cleaned up. At the same time, the program has become a bureaucratic nightmare, consuming more than 20 percent of the EPA's $6.8 billion budget. Moreover, 47 percent of the Department of Justice's enforcement actions for major environmental programs are dedicated to the Superfund program. From 1995 to 1997, a total of 839,500 hours were billed to Superfund enforcement, according to the Department of Justice. Without fundamental reforms, the costs will only increase. A study conducted at the University of Tennessee estimates that, under the present system, cleaning up 3,000 Superfund sites will cost between $150 billion and $352 billion.

"Superfund is a flawed program built from a premise that guarantees its failure."

Despite the dismal record of the Superfund program, the [former] Clinton administration and [former] EPA Administrator Carol Browner defend the program, claiming that administrative reforms have effectively fine-tuned the program and hastened the pace of cleanups. As Browner recently stated, "By any measure, we are making a great deal of progress in our efforts to improve the nation's hazardous waste cleanup program—to make it faster, fairer, and more efficient—and to ensure that it does the best possible job of protecting the health of our citizens and returning land to communities for productive use."

The Cleanup Process

Superfund is a comprehensive program that starts with the identification of potential hazardous waste sites with continued oversight through the final remediation, or clean-up, of priority sites. To identify sites, the EPA has created an extensive database known as CERCLIS (Comprehensive Environmental Response, Cleanup, and Liability Information System), which catalogs potential hazardous waste sites based on information provided by the public and the EPA's internal studies.

Once a site is catalogued in CERCLIS, a preliminary assessment determines whether further action is necessary. Where further action is warranted, the EPA will conduct a site inspection to assess the potential risk and what remediation may be required. At this point, about one-third of the sites are listed as requiring no further action. For the remaining sites, the EPA uses its Hazardous Ranking System (HRS) to evaluate the severity of the problem. A ranking of 28.5 or higher qualifies a site to be placed on the National Priorities List (NPL), which qualifies a site for federal funding. Although the NPL is supposed to ensure that the most dangerous health risks are addressed most swiftly, critics of Superfund claim that NPL status does not necessarily correlate with health risk. Some sites on the list may pose little or no health threat, while others that are not listed may actually pose greater health risks. The HRS ranking of 28.5 is arbitrary, a political artifact of the original legislation that required 400 sites to be listed on the NPL (with at least one in every state). Today, there are 1,192 sites on the NPL, and the EPA has proposed adding another 52 sites.

"Critics of Superfund claim that [sites given priority for clean-up] . . . may pose little or no health threat."

Two basic cleanup options are available to the EPA under Superfund: short-term removal actions and long-term remediation. Removal actions spark little debate and are widely acknowledged to be effective in removing immediate health risks. Often, removal actions are emergency measures aimed at cleaning up chemical spills and removing hazardous wastes that threaten a local population.

Long-term remediation efforts are far more controversial, and questions about their funding and procedures lie at the core of the debate over Superfund reform. Once a site is listed on the NPL, the cleanup process involves significant oversight by state and EPA officials. The process begins with a remedial investigation to determine the physical and chemical properties of the site and the exposure levels of the local population. The EPA regional administrator is responsible for evaluating the findings and selecting the appropriate method of remediation, which is referred to as the Record of Decision (ROD).

> ***"Strict liability means that individuals can be held responsible regardless of intent or negligence."***

To comply with the ROD, those responsible for the cleanup develop a remedial design, which will then be used to carry out the final remedial action. At large Superfund sites, this process can be lengthy because the remediation process is sub-divided into a number of individual operating units, each of which requires its own studies and cleanups. Once the remedial action is completed, the EPA, in conjunction with state officials, can remove the site from the NPL. Due to the fact that sites often must be maintained and monitored over time, sites commonly remain on the NPL but are shifted to the "construction completed" category, where many sites remain for years before being removed from the NPL. The EPA currently lists 506 sites as construction completed and 162 sites as deleted from the NPL. However, these numbers do not indicate how many cleanups actually have occurred. For example, a study by the General Accounting Office (GAO) of 149 sites found that for 21 percent of the sites no cleanup was necessary.

Tag, You're It!

The liability system that serves as the foundation for the Superfund program is as expansive as it is flawed. For the purpose of Superfund cleanups, liability is strict, retroactive, and joint and several. Strict liability means that individuals can be held responsible regardless of intent or negligence. Retroactive liability means that individuals can be held responsible for past actions, even if those actions were legal at the time. Finally, joint and several liability means that any given individual may be held liable for the full cleanup, not just a proportionate share. Taken together, these liability rules allow the EPA to cast a wide net for PRPs and deep pockets that can bear the costs of the cleanup. Due diligence and other standard legal defenses do nothing to minimize the threat of being identified as a PRP. The original CERCLA legislation identifies four specific groups to whom the liability standards apply: present owners and operators of a site, past owners, individuals who generated the waste at a site, and transporters who hauled the waste to the site.

Only orphaned sites are cleaned up using the trust fund. In all other instances, the EPA relies on the Superfund liability scheme to identify PRPs that will pay for the necessary remediation. The EPA can either begin the remediation process itself, using money from the trust fund, or the agency can issue an administrative order demanding a PRP to begin the cleanup process under the threat of fines that start at $25,000 per day and can rise to $75,000 per day for ensuing violations. Either way, PRPs will eventually pay; if the EPA conducts the cleanup itself, it will seek reimbursement from PRPs.

When the EPA identifies PRPs and demands their participation in the remediation process, the cogs of the legal system begin churning. Existing PRPs begin filing "contribution suits" to expand the number of PRPs involved in order to reduce their share of the cleanup costs. The list of PRPs quickly expands from the handful of "corporate polluters" identified by the EPA to a long list of smaller and smaller businesses such as "mom and pop" businesses on Main Street, hospitals, the Girl Scouts, and even an Elks Club in one instance.

To understand the reach and tangle of Superfund, consider Petroleum Products in Pembroke, Florida. This was a recycling site where people sent waste oil to be recycled. The site was placed on the NPL in 1987, and somewhere between 2,000 to 4,000 PRPs have been identified. The PRPs include cities, the University of Miami, the Miami Fire Department, and thousands of others who thought they were acting responsibly by sending their waste oil to a recycling facility. Although many have been dragged into the liability maze, no long-term cleanup has been done in the ten years that the site has been on the NPL.

In another instance, the original PRPs at a site in Ludlow, New York, called a meeting at the local auditorium, inviting all the local businesses to attend. At the meeting, the small businesses were told that they should offer a payment to cover part of the cleanup or else they would be sued. The small businesses decided to pay $1,500 each to get out of the Superfund web, because it would be cheaper to pay than go to court. These payments were made under the threat of litigation, rather than as a result of any evidence that found them liable for the site. As Rep. Sherwood Boehlert (R-NY) stated: "They were operating under the current Superfund law. We should never let this happen again."

Clearly, leaving PRPs to sort out their differences in court can be a costly venture, as expenses are compounded by enforcement actions and lengthy negotiations with the EPA at every step of the remediation process. Moreover, the legal and other transaction costs incurred contribute nothing to the remediation process. Numerous studies have examined the burden of transaction costs on the Superfund program. As noted earlier, RAND estimated that up to 36 cents out of every dollar goes to transaction costs instead of cleanup. And the greater the number of PRPs, the higher the transaction costs. Another RAND study found that at single party sites, transaction costs were 5 percent of total costs with costs rising to between 34 percent and 36 percent of total costs at sites with more than 16 PRPs.

"Up to 36 cents out of every dollar goes to transaction costs instead of cleanup."

The liability system's legacy is also evident when examining the length of time required to clean up Superfund sites. Fifty-three percent of the orphan sites, where no PRPs are present and cleanup is funded by the EPA, have been cleaned up. At sites with more than 500 PRPs, however, only 17 percent have been cleaned up.

Despite mounting evidence that the liability system diverts resources away from cleanups and delays their completion, the EPA continues to defend the liability system on the grounds that it embodies the "polluter pays" principle. Any changes to the current liability system, they claim, will "let polluters off the hook." A closer examination of the Superfund, however, demonstrates the weakness of this claim. First, the simple scope of PRPs should suggest that the liability system hews too broadly and indiscriminately to be seen as punishing polluters. In fact, an argument could be made that the liability system provides polluters with incentives to shift the burden to those who bear little or no responsibility for a Superfund site. To force a small restauranteur to pay tens of thousands of dollars in cleanup costs for sending mashed potatoes to the local dump is hardly taking a stance against polluters.

Also, the Superfund itself is funded by a special tax on those identified by the EPA as corporate polluters—oil and chemical companies. Such companies are forced to pay into the trust fund, irrespective of their records as environmental stewards. Even in the absence of identifiable PRPs, "polluters" are paying. Unfortunately, this tax does little to provide incentives for favorable environmental practices; the tax makes no distinction between good and bad actors.

Another claim made by supporters of the status quo is that without the threat of the Superfund program, bad actors will continue to pollute. Yet, it is important to remember that Superfund was created to address existing hazardous waste sites, not to regulate current waste management practices. There are numerous federal, state, and local laws and regulations that govern the disposal and management of hazardous waste. Most significant is the Resource Conservation and Recovery Act (RCRA), which was passed in 1978 to regulate treatment, storage, and disposal facilities. Also, the legal system may be used where parties have disputes over land use or dangerous activities. . . .

What Are We Cleaning?

In addition to the liability quagmire, the remediation process itself has been a constant source of criticism for the Superfund program. Not only are there concerns over excessive administrative costs and poor management practices, but oftentimes the cleanups are costly and bear no relation to eliminating identifiable health risks for individuals in the local community.

The process used by the EPA to determine the level of remediation at a given Superfund site begins with a risk assessment in which the agency uses a number of assumptions to determine the level of risk and the exposure of the local population. However, the EPA has typically relied on what many claim are unrealistic assumptions and conjectures about the future use of the site in question. Excessively conservative assumptions tend to overstate the risk and raise remediation costs significantly. Currently, the average cleanup costs more than $30 million.

For example, consider the following EPA standard for cleanup, "For 350 days per year a child is assumed to eat 200 milligrams of dirt while playing in the soil at a Superfund site that may be surrounded by fences with few residences nearby." Such demands for high-quality, edible dirt are made in instances where there are few residential areas and where a fence to keep trespassers out surrounds the Superfund site. Highly speculative conjectures about future use allow the EPA to demand expensive remediation programs to eliminate any potential pathways of exposure. It is often the assumption that in the future a residential neighborhood may take over an industrial site that raises the EPA's assessment of potential health threats. A study of 77 Superfund sites found that 91 percent of the cancer risk claimed by the EPA was future risk that depended on the EPA's assumption of future land use and behavior.

> ***"Oftentimes the cleanups are costly and bear no relation to eliminating identifiable health risks."***

Clearly, any meaningful reform of Superfund must address risk assessments and remediation technologies required by the EPA. Costly decisions have been made in the past because Superfund has been viewed as a cleanup program, not a risk reduction program. Decisions about which sites move forward have been made based on administrative requirements (such as the length of time a site has been on the NPL) rather than risk assessments that identify real hazards to public health. Sound science and more accurate risk assessments can significantly reduce the costs of remediation, while reducing real health risks when they are found. Steven J. Milloy of the National Environmental Policy Institute suggests that the costs of cleanups would fall by 60 percent if the program focused more directly on risk when identifying the appropriate remedies. At the same time, resources would be allocated to those sites posing the greatest health risks.

Is the Program Getting Better?

In response to proponents of a major overhaul of the Superfund program, the EPA moved forward with a number of administrative reforms to increase the pace of cleanups. The agency points to the fact that 268 sites have been cleaned up in the past four years, which is more than the previous 12 years combined. In conjunction with the administrative reforms, the EPA proposes moderate Superfund legislation, claiming that the EPA will "hold firm to the 'polluter pays' principle and to ensuring that money is going to cleanups, not to lawyers." These goals appear to contradict each other, because it is difficult to keep the money away from lawyers when it is a liability-based system. And a closer look at the numbers suggests more than moderate reforms are necessary.

In fact, there is little data to suggest that the program is working better today than it has in the past. A GAO study of the EPA's administrative reforms found that out of 45 reforms only 6 could be said to have quantifiable accomplishments, and only three had documented accomplishments. As David Aylward, president of National Strategies, Inc. (NSI), testified before Congress: " [T]here is no evidence that the program has been significantly improved, and current data indicates that the program has worsened, at least in outputs." Mr. Aylward bases this assertion on analysis of the EPA's data conducted by NSI that provides a careful assessment of Superfund's progress. As for the EPA's claim that more sites are being categorized as construction complete, the NSI research finds that this is simply a function of time. Given that most sites were listed in the mid-1980s and that cleanups have been averaging 10 to 12 years to complete, it follows that they would be finished, or close to finished, in recent years. Consequently, the current pace of completion simply confirms earlier estimates that cleanups require at least 10 years.

The NSI analysis goes on to show that three other measures of accomplishment also declined from 1996 to 1997: remedial cleanups completed, remedial designs completed, and remedial actions started.

It is not surprising that the administrative reforms have not altered the program significantly. Administrative changes at the margin do not alter the underlying principles that drive the Superfund machine. Raising funds through the liability system continues and parties drawn into a Superfund site continue to file contribution suits and drag others into the maelstrom. Legal maneuvering diverts resources away from questions of public health while delaying cleanups. Until these core issues are addressed, reform can only achieve minor successes at best.

Meaningful Superfund Reform

Superfund is in clear need of fundamental reform. Good intentions have wrought a bad program that is broken and has not served the public well. Even EPA Administrator Carol Browner admits: "[W]e've got to have legislation. And we are 100 percent committed to enacting responsible Superfund legislation." Unfortunately, for the EPA this means focusing almost exclusively on "getting the little guys out." Real reform must go further. Both the scope of the program and the remediation process must be recast to promote swifter and more efficient cleanups that focus on real threats to public health.

> ***"There is little data to suggest that [Superfund] is working better today than it has in the past."***

Hazardous waste sites, in fact, are local problems and should be addressed at the state and local levels. The federal National Priorities List should be capped, with future site identification and remediation conducted by state and local governments. Already, a number of states are moving forward with their own cleanup programs. Michigan, for example, employs over 400 people and has spent close to $1 billion on cleanups. State level programs need the flexibility to be innovative in their approach to remediation. Simply transferring the federal program—with its burdensome remediation process and costly liability system—down to the state level will do little to expedite the cleanup process. Allowing states to develop their own approach to site identification and remediation will foster cleanups that are suited to the specific circumstances of different states. In addition, states could work to privatize Superfund sites, whereby states could auction off waste sites. The price could be positive (if cleanup costs are lower than the value of the site) or negative (if cleanup costs are greater than the value of the site). With a negative price, the state would pay the bidder to clean the site. In exchange for the rights to the property, the winning bidder would post a bond that the government would hold until the health risks have been mitigated.

> ***"Superfund is in clear need of fundamental reform."***

This would allow the Superfund program to focus exclusively on remediation at the NPL sites. Improved, risk-based remedy selection and a streamlined administrative process would reduce the costs of the remaining NPL cleanups. These savings and the finite scope of the program would eliminate the need for new taxes. This is especially true if the EPA meets its stated goal of cleaning another 400 sites by the end of the year 2000. If the EPA required additional resources to complete the remediation of NPL's sites, funding should be allocated from the annual budgetary appropriation to the EPA.

Remedy reform alone, however, will not eliminate the inherent inefficiencies of Superfund. Reform must address the liability system, which has proven to be a stifling, inefficient way to raise funding for Superfund cleanups. The current liability system bears little semblance to the "polluter pays" principle, nor does it provide incentives for environmental responsibility. Repealing retroactive liability would provide the greatest return in terms of eliminating transaction costs and promoting cleanups; identifying ways to minimize the applicability of retroactivity may be the second-best solution. Examples include exemptions for generators and transporters who were not bad actors and did not have responsibility over site management, exemptions for co-disposal sites (which typically have numerous PRPs and poor record keeping, making it difficult to allocate responsibility), or exemptions for small businesses. What is clear, however, is that reforms must be broader than simply exempting the "little guys." There are too many sites where the PRPs are so numerous that even if the *de micromis* and *de minimis* [parties who made a good faith effort to comply with the law] were exempted, the legal wranglings of the remaining PRPs would still delay the program.

Reforming joint and several liability would also significantly reduce the legal proceedings that delay the cleanup process. Making a PRP potentially responsible for the full cleanup, no matter how slight the PRP's involvement, generates incentives to file contribution suits and expand the pool of PRPs. Clarifying that liability means liability for a PRP's proportionate responsibility ensures fairness while reducing the need to litigate.

Without significant reform, hazardous waste cleanups will remain mired in legal battles. Changes at the margin or new layers added to the existing Superfund program offer little hope. Superfund must be reformed substantively and structurally. States should play a greater role in cleaning up waste sites, remedy selection must be based on risk assessment and sound science, and excessive oversight costs must be reduced. The original structure of the program must be revisited, with an eye toward extricating the program from the courtroom. Until then, Superfund will remain as an example of failed environmental policy.

Clean Water Regulations Are Ineffective

by Richard A. Halpern

About the author: *Richard A. Halpern is the director of environmental affairs for the Hudson Institute's Center for Global Food Issues. The Hudson Institute is a conservative public policy research organization with offices in Indianapolis and Washington, D.C.*

In 2000, when [former] Environmental Protection Agency (EPA) administrator Carol Browner was refining a sweeping new clean-water rule, she said that it would finally enable us, twenty-eight years after passage of the Clean Water Act, to "finish the job of restoring the nation's waters"—in another fifteen or twenty-five years. The environmental groups pushing for the rules were furious and fired off an angry letter to Browner: "This delay is both unconscionable and contrary to law." They argued that in 1972 Congress intended "that water-quality standards be attained within twelve years. It is now twenty-six years [since this new rule] . . . was to have been in place, and fifteen years after the goal of clean water was to have been met." This angry response by six powerful members of the environmental establishment ended with a threat to withdraw their support from the EPA.

A Regulatory Fraud

To put it bluntly, the Clean Water Act (CWA) of 1972—acclaimed as "the American environmental movement's single most important achievement"—isn't up to the job. In fact, to put it *very* bluntly, the CWA is something of a fraud. More accurately known as the Federal Water Pollution Control Act, this law set in place a regulatory system that does not and cannot promise verifiable environmental benefits. Instead of pinpointing problems and dealing with them directly and efficiently, it has been striking out blindly for almost thirty years, attacking with equal force both imagined and real water-quality problems, sometimes hitting, sometimes missing, but always gobbling up billions of dol-

From "1491 and All That," by Richard A. Halpern, *American Outlook*, November/December 2000.

lars at a gulp. The Clean Water Act was the equivalent of legislating bypass surgery for everyone in the country because someone near Cleveland died of a heart attack. It forced expensive wastewater treatment technologies on everyone everywhere, regardless of their actual water quality. It was indiscriminate, but it was evenhanded, falling, like the rain, on the clean and the unclean alike. The price tag for the first twenty years was $540 billion.

With the Clean Water Act, Congress told the nation that the waters could be made clean, without a rigorous scientific evaluation, simply by installing technology to prevent anything harmful from getting into our waters. The preferred option was zero discharge. As the EPA's first administrator, William Ruckelshaus, cautioned Congress, however, in 1971 testimony before the House Committee on Public Works, zero discharge would replace the goal of clean water with a goal of implementing technologies that might or might not give us clean water; we would never know whether they were successful. "The social benefit we are all seeking—high water quality—is eliminated from the equation," he told the committee. Committee chairman John Blatnik (D–MN) warned his colleagues that "the technology for 'no discharge,' at least that which is . . . feasible, does not now exist."

Child of Panic

The Clean Water Act was a child of panic, formed in a climate generated by television pictures of the 1969 Cuyahoga River fire near Cleveland and images on the evening news of naked sewer pipes discharging presumably noxious liquids into various bays, lakes, and rivers around the country. In 1970, as environmental writer Ronald Bailey reminded us in a May 2000 Earth Day essay for *Reason* magazine, the first Earth Day produced "a torrent of apocalyptic predictions." Ecologist Kenneth Watt warned that we had but five years to stave off Doomsday. Harvard biologist George Wald estimated that civilization would only survive another fifteen or thirty years unless drastic action was taken. Others predicted that mass starvation, in which four billion would perish, was only a few years away.

In keeping with the eschatology of the day, Alabama congressman Robert Jones urged passage of the act: "If we wait too long, all the dollars on earth won't buy back what we have lost. . . . We may well destroy all hope of saving our environment." In the Senate, the bill's chief architect, Edmund Muskie (D-ME), postulated "a threat to life that could not be more real . . . could not be more desperate . . . grim realities of lakes, rivers, and bays where all forms of life have been smothered . . . and oceans which no longer provide us with food." The "cancer of water pollution," he said, "threatens our very existence." Ruckelshaus, by contrast,

> ***"[The Clean Water Act] set in place a regulatory system that does not . . . promise verifiable environmental benefits."***

attempted to put the "grim reality" into perspective. "The fact of the matter is that in the country it is 10 percent of the dischargers causing 90 percent of the problem." Both he and Russell Train, his eventual successor at EPA, urged Congress not to short-circuit the fundamental logic of problem solving but to determine the nature and scope of the water-quality problem, analyze the causes, and apply appropriate remedies to specific, documented deficiencies.

"The Clean Water Act was a child of panic, formed in a climate generated by television pictures of the 1969 Cuyahoga River fire."

Unfortunately, intensifying the sense of crisis and mood of apprehension in Congress was what the Senate Public Works Committee Called the "information gap." No one had ever collected the data needed to define the problem. The nation's waters were unmonitored, and because no two waterways are truly comparable and no single water body is exactly like itself for two miles, two years, or even two months, site-specific monitoring over time would be indispensable in establishing standards. Without credible baselines and with little understanding of natural background conditions, no one really knew what real-world water quality was or should be. Even worse, there was no way to demonstrate convincingly the impact a discharge had on the quality of the water receiving it. As New York Congresswoman Bella Abzug told Train, "You can't tell who is causing the problem." Train and Ruckelshaus begged for more time to resolve the scientific issues, but Congress, trapped in its own rhetoric of apocalypse, rejected their approach as unworkable. It abandoned the attempt to establish empirically based water-quality standards or even to define its terms in scientifically meaningful ways, putting off those crucial tasks to a future that would never arrive.

Unrealistic Goals

Congress said that the purpose of the CWA was "to restore and maintain the chemical, physical, and biological integrity of the Nation's waters," but it did not define "integrity." That was to be accomplished later, as Muskie told Senator Bob Dole (R-KS), through a "nationwide program of research . . . designed to accumulate, as quickly as possible, the vast amount of information which we still do not have." The CWA set a national goal of achieving "zero discharge"—the complete cessation of discharge of pollutants—by 1985. But even this boldly stated deadline was vaporous. When Senator James Buckley (D-NY) challenged the goal as unrealistic, "something . . . there is every reason to believe cannot be achieved by the year 1985," he was given Muskie's assurance that the 1985 deadline was only "a policy objective . . . not locked in concrete. It is not enforceable."

The list of substances the CWA defined as "pollutants" was exhaustive: "The term 'pollutant,' means . . . sewage, garbage, chemical wastes, biological materials . . . rock, sand . . . industrial, municipal, and agricultural waste." The law might just as well have said "biological and non-biological materials" and left it at that. Under the CWA, it became illegal to discharge virtually any substance into the waters of the United States.

Some members of Congress, including Buckley, wondered whether the act intended to outlaw all impact on the nation's waters resulting from human activity. As it turned out, Congress had indeed defined "pollution," as "the manmade or man-induced alteration of the chemical, physical, biological, and radiological integrity of water." Anything humans did that affected water could be considered "pollution." Buckley asked the senate leadership whether Congress was adopting "the 1491 Standard" for water quality—a restoration of pre-Columbian conditions. On the Senate floor, Tom Eagleton (D-MO) agreed that "it would be impossible . . . to restore the temperatures of our rivers without reforesting all the banks of all the tributary streams." What then would "integrity of the nation's waters" mean? The concept would not be defined by the EPA, Eagleton declared, but by the states. And the states, he added, would have one year to perform the (impossible) task of accumulating and evaluating that vast amount of information that Congress acknowledged no one had.

Lack of Integrity

Twenty-eight years later, we are still waiting for that information. In congressional hearings in 2000, the Association of Metropolitan Sewerage Agencies (AMSA) revealed that many water-quality standards (the states' definitions of "integrity") were, in essence, "wish lists" established in the 1960s and 1970s without significant scientific analysis. Officials are finding out, AMSA reported, that in many cases these standards "don't make any sense."

When it passed the act in 1972, Congress said, quite accurately, that it didn't know the true condition of our waters. The mandates that waters be inventoried and standards set grew out of precisely this complete dearth of knowledge. And this year's congressional hearings on Browner's new rules made it all too clear that we don't know very much more about their condition now, nearly thirty years later. We don't know what our quality was then, what it was in 1491, or what it is now. Hence, the conventional wisdom on the Clean Water Act—that the legislation has dramatically restored to health two-thirds of the nation's waters—is a fabrication, a complete fiction.

> ***"Under the [Clean Water Act], it became illegal to discharge virtually any substance into the waters of the United States."***

The EPA is now telling us that 40 percent of America's waters are unacceptably polluted, that 218 million Americans—nearly 80 percent of us—live within ten miles of polluted waters. The EPA's claims, however, are based on water-quality reports from the states, and the states admit, in a survey conducted by the U.S. General Accounting Office

(GAO) in spring 2000, that their reports are a sham. They are based on outdated assessments, anecdotal evidence, and random observations, not scientific sampling. Only one state claims to have significantly more than half the data it needs to make a sound assessment of its waters, and only 10 percent of the nation's river miles have ever been monitored, even once. In fact, some states report that the EPA encourages them to pad their list of polluted waters. Since federal dollars follow listed impairments, such padding may, at times, require little encouragement.

> ***"The conventional wisdom on the Clean Water Act—that [it] has dramatically restored . . . the nation's waters—is a fabrication."***

The information gap has some truly debilitating consequences. Attempting an analysis of water quality in the Mississippi River Basin for a 1992 report to EPA, Purdue University researcher Steven Lovejoy encountered major problems finding data. He found that "measurements of actual concentrations of pollutants and flows were nonexistent." In desperation, Lovejoy sampled key locations during 1989. Without baseline data, however, he could not ascertain whether 1989 was an average year. Conclusions based on such data have little validity and less value. Yet even data as impoverished as these are routinely reported as "best available data." As such they become the basis for critical environmental policy decisions affecting the lives of millions of Americans.

A Pretext for More Regulation

After nearly thirty years and $600 billion worth of hit-and-miss technologies, we still don't know what has been achieved or what still needs to be done. Also, the lack of real-world data leaves activists free to claim that "our water continues to be poisoned," and provides them a pretext for demanding an increasingly intrusive role in determining the national lifestyle. *The Wall Street Journal,* commenting on Browner's new rule, observed, "Environmental policy has been all but separated from scientific evidence and is now achieved through administrative fiat to sate activists who themselves can no longer be bothered with letting science get in the way of what they want." This is all a direct legacy of Congress's 1972 inability to define its terms according to any "rule of reason" (in Democratic Texas Senator Lloyd Bentsen's phrase), its decision to seek enhancement of the nation's water quality without sound science, and the vagaries and vacuities of the criteria and deadlines outlined in the CWA. Thus it should not surprise us that the CWA has occasioned endless litigation. As Ruckelshaus observed in 1996 regarding the CWA and several other environmental statutes, setting out on "the Mission Impossible of pursuing perfection leads directly to the devolution of all important environmental decisions to the courts."

The only way to ensure the quality of our nation's water is to commit the necessary resources to gathering sufficient information for a sound water-quality program grounded in science. Since 1972, the United States has spent close to $600 billion on wastewater treatment technologies. We have spent more than $1 trillion on environmental protection in the last eight years alone, but, according to a 1990 estimate, we spend just $33 million a year to monitor our waters. A true and complete national water-quality inventory will take a full twenty years just to make a proper start. The GAO's survey revealed not only that state officials lack good data but, that they are confident—like the homeowner who knows which windows need caulking—that they know where the major problems are and how to address them. There is one way to find out if they are right.

Clean Air Regulations Are Ineffective

by Richard L. Stroup

About the author: *Richard L. Stroup is a professor of economics at Montana State University and serves on the board of advisors of the Independent Institute, a conservative think tank based in Oakland, California.*

In 1990, Congress passed the Clean Air Act Amendments (CAAA). Title III of the Act required the Environmental Protection Agency (EPA) to regulate emissions of toxic or hazardous pollutants. This controversial Title was the culmination of pressure by environmentalists to impose more stringent controls on airborne emissions of 189 potentially toxic airborne substances in addition to the "criteria" pollutants such as sulfur dioxide, carbon monoxide, and ozone. About 250 major industrial emitters of the newly listed substances, including chemical plants and oil refineries, would have to install "maximum achievable control technology" (MACT).

High Costs, Low Risk

There are two major problems with Title III. The first is that the available evidence indicates that the costs greatly outweigh the benefits. In fact, the evidence suggests that while the costs of added controls are large, there may be little or no benefit from the added regulation. The strongest claim to a benefit from Title III is that it may reduce the risk of getting cancer. But the risk from these pollutants is uncertain, and now appears to be smaller than risks that we accept every day—risks caused by others without the consent of the person at risk, just as any risks from air toxics would be for people who lived near the toxic emissions. In contrast, the costs of the regulation are substantial and are directly paid by a relatively small number of industrial firms.

The concentration of large costs raises a second major problem: that of fairness. A few firms are being made liable for very costly control measures even though there is little evidence that they have done anything wrong or that the

Excerpted from *Cutting Green Tape: Toxic Pollutants, Environmental Regulation and the Law*, by Richard L. Stroup, edited by Richard L. Stroup and Roger E. Meiners (New Brunswick, NJ: Transaction Publishers, 2000).

emissions to be controlled at such great cost are actually imposing (or would impose) significant harm on anyone or on the environment. To impose such costs to avoid unconscionable harm on even one person might be perfectly just. But no such harms or risks are demonstrable. . . .

Pollution control is special as a policy issue only because of the knowledge problem. If a particular pollutant is released at a specified time and place, who is (or would be) harmed, and by how much? If we could easily learn the answer to that question for any given (or contemplated) release, a simple solution would be available: Forbid those that would be truly harmful, just as we stop personal assaults, and for the same reason—imposing such harm is wrong. In fact, common law has for many decades offered this sort of protection. A plaintiff who can show that he suffers real and substantial harm (or is about to be harmed) by an emitter of air toxics can generally get court-ordered relief in the form of damage payments from the polluter and/or an injunction to stop the emissions.

"Evidence suggests that while the costs of added [air pollution] controls are large, there may be little or no benefit from the added regulation."

With full knowledge of this sort, potential polluters know that harmful emissions will not be tolerated and will find other times and places for their releases, making them less harmful and thus permissible. Or they will simply find other, nonpolluting ways to achieve their goals just as would-be perpetrators of assault refrain from using assault to get what they want when a court, knowing who did what to whom, would discipline them.

The problem, of course, both in the case of polluters and muggers, is in showing who has done what to whom. Has the accused polluter or mugger actually caused harm? Often the knowledge about what effects pollution is having, or will have, is difficult to obtain. The common law will not punish defendants for alleged harms if plaintiffs cannot show by weight of the evidence that the defendants caused those harms. So instead, possible victims of pollution turn to regulation precisely because the information available to judges and juries about pollution and its effects is so sparse. Unfortunately, turning to regulators does not provide the information that is needed for rational pollution control: how harmful is the pollution in various situations?

Information is not the only requirement for proper control. There is also the question of what constitutes harm. Even the clear presence of some degree of unwanted activity should not automatically trigger enforcement activity by government. For example, if I run to get on a train and accidentally knock you down, but then apologize and help you up, my "assault" will probably go unpunished so long as there is no intent to harm and no damage. Some tolerance is expected, in any society, when incidental damage is below some threshold, or *de minimis* level.

Determining Acceptable Risk

How do we decide what is a *de minimis* outcome or air pollution risk? One revealing way to think about this question is to ask whether the public accepts risks in other circumstances that are comparable to the risk from air toxics. Bernard D. Goldstein et al. offer data on an involuntary risk that is comparable to harm from airborne toxins: the risk of death due to an airborne object—a falling airplane—faced by people on the ground. Their statistics show that for a seventy-year lifetime, the risk of death from that source is 4.2 per million. Today, in other words, an average American citizen who otherwise would live to age seventy is exposed to a 4.2 per million chance of being killed by a falling airplane over her lifetime. This danger is very small, and appears to be acceptable to U.S. citizens. How do we know that? One indicator is that there is little pressure to decrease the risk by sharply upgrading regulatory requirements for airplane maintenance and pilot training or by, say, banning pleasure flights. Arguably most existing regulations were put in place to protect passengers and crew, rather than groundlings. Perhaps somewhat greater risks of air crashes would be tolerated if only the groundlings' safety were at stake, rather than the safety of other, additional people at the same time. A second indicator of the risk's acceptance is that people living near airports seem not to take any special precautions.

How does the 4.2 per million risk of death compare to the equally involuntary risks to neighboring citizens from the air toxics emissions regulated under Title III of the Clean Air Act Amendments? Title III has two general requirements for the 189 listed air pollutants. The first ignores risk entirely, requiring polluters to use Maximum Achievable Control Technology (MACT), regardless of actual risks from the specific emissions to be controlled. The MACT requirement is expected to bring any risk to the most exposed resident of the area down to less than one per million—one-quarter the risk of death from a falling airplane. The second requirement is that if the residual risk exceeds that level, in the judgment of the EPS, the EPA Administrator must promulgate standards to further reduce the remaining risks. Most pollution standards are driven either by technology requirements, such as MACT, or by a requirement to keep the risk of harm below stated risk levels. Title III imposes both constraints on emission levels for the listed air toxics. But as we will see below, the EPA calculates expected risks from any facility by using procedures that are expected to far overstate the risks. Let us look more closely at what is known about the risk of cancer from airborne toxic emissions, and how that information is utilized by the EPA.

Overestimating the Hazards

The EPA has estimated that the leading hazardous air pollutants (about 100 in total) may cause between 1,700 and 2,700 additional cancer cases per year. But outside analysts disagree. Paul Portney of Resources for the Future says that the

EPA's risk assessment methodology is "designed to overestimate risk far more often than it underestimates it," and he estimates that 500 cases is "probably a generous estimate of the reduced cancer risk associated with the emissions controls." In fact, he estimates that benefits may be as low as zero, because it is possible that no cancers will actually be prevented. His "most likely" estimate is that the benefits will be worth $1 billion, still far less than the cost of the control technologies required under Title III of the Act, projected to run from $6 to $10 billion per year.

John Graham, of the Harvard School of Public Health, has also been skeptical. He testified before a congressional committee in 1989 that the daily inhaled levels of air toxics are several orders of magnitude (several multiples of ten) smaller than levels studied by scientists. "There is no direct evidence that outdoor exposure to toxic air pollutants is responsible for a significant fraction of disease or mortality. . . . EPA's widely quoted cancer risk estimates should be treated with caution because they are based on some questionable assumptions—for example, that humans are as sensitive as the most sensitive tested animal species, and that any human exposure to a carcinogen, no matter how small, results in some increase in cancer risk." These assumptions and many others are made by the EPA in order to make their risk assessments conservative—that is, more likely to overstate than to understate the true risk. Often the overstatement is severe.

As an illustration of the EPA's conservative assumptions and their effects, consider the EPA estimates of the risk of cancer caused by airborne emissions from coke ovens, which make coke from coal for use as a fuel in steel-making. The risk assessment process has been carefully critiqued by Frederick Reuter in an industry-sponsored consultant's report to the National Research Council Committee on Risk Assessment of Hazardous Air Pollutants.

The EPA cites studies of steel workers inside coke-making plants as having revealed, as summarized by Reuter, "coherent patterns of excess relative risks of death due to respiratory system cancer, relating directly to both the level and the duration of exposure to coke oven emissions. . . ." The EPA used this to justify costly control measures to protect people who lived nearby. Using primarily data from the EPA, Reuter pointed out that workers in coke plants who were not working close to the ovens, and thus had much lower exposure to the emissions, showed "no coherent patterns of excess relative cancer risks." In fact, the studies utilized by the EPA showed that among lower-exposure workers at coke plants, those with less exposure actually had a greater level of cancer mortality. Apart from the high-exposure workers located at the coke ovens, the studies showed no apparent added respiratory cancer risk from working in coke plants, even though the exposure to emissions there was much larger than in the surrounding neighborhoods.

"There is no direct evidence that outdoor exposure to toxic air pollutants is responsible for a significant fraction of disease or mortality."

Given the absence of evidence that coke oven emissions cause off-property cancer risks, how could the EPA justify its demand for expensive controls to reduce further these very low neighborhood exposures? The agency's risk assessments show that for maximally exposed individuals outside the plant, a risk estimate can be constructed that will suggest the possibility of an unacceptable risk.

False Assumptions

In the absence of good data on human exposures outside the plant to air toxics emissions, on the carcinogenicity of calculated exposures, and on the other components needed to assess the resulting risk, the EPA fills in data gaps with a number of assumptions, each tending to overstate the health risks. The resulting upwardly biased numbers are typically multiplied together, resulting in a compounding effect that is likely to further overstate the risk. Reuter's report provides details on the product of the effects of EPA assumptions, each of which is constructed to bias the risk assessments upward, in order to avoid possible understatement of the risks. (Use of unbiased estimators, if the EPA had done so, would have resulted in estimates equally likely to understate or overstate the true risk.) The biased estimators that the EPA actually used can be hundreds, thousands, or tens of thousands of times higher than the unbiased or "maximum likelihood" estimators.

> ***"The [EPA] fills in data gaps with a number of assumptions, each tending to overstate the health risks [of air pollution]."***

Reuter uses EPA data and risk assessments at the Clairton, Pennsylvania, coke-making plants of USX Co. to show the degree of overstatement when several conservative assumptions are used in place of actual data on the various components of EPA risk assessments. EPA models used to assess the cancer risk from coke oven emissions to the hypothetical "Maximally Exposed Individual" (MEI) supposedly living at the edge of the plant's property, yield a strongly biased risk assessment. Reuter's study uses primarily EPA data and some more sophisticated EPA models to challenge the validity of those risk assessment findings. . . .

The EPA had calculated that without additional controls cancer deaths caused by the Clairton facility would amount to 2.83 per hundred people, indicating that a person so exposed would have a risk of dying from cancer that is increased by more than one chance in forty. Instead, Reuter calculates from much the same data base that the best estimate of increased risk of death from cancer is not one in forty persons, but about one in 10,000,000 persons.

As calculated by Reuter, the risk to a person on the ground of death from an airplane falling from the sky is roughly 250,000 times greater than the risk of dying from cancer from maximum exposure to coke oven emissions. If Reuter is correct, the Clairton risk might best be viewed as *de minimis,* or well below the threshold level of risk that Americans find tolerable, since the far greater risk from the falling airplane is easily accepted by regulators and by the public. In trying to be so protective by distorting specific risks, the EPA may well distort resource use so much as to reduce incomes sufficiently to cause more, rather than fewer health problems. The recent literature on this general topic supports that hypothesis.

The Extreme Costs of Regulation

If the Clairton controls are representative, then Title III appears to cause a massive waste of resources, since the expenditures represent real resources devoted to actions that in many cases are probably useless. It seems quite likely that at many facilities, congressional assumptions that health risks are present that justify the imposition of MACT standards are false. Similarly, it seems likely that EPA's risk assessments exaggerate estimated risks so much that their findings will often be invalid. Even if the risks are real, a far simpler, less disruptive and less costly solution could be adopted. For example, the polluter could be allowed to try to purchase the land on which exposures otherwise would occur and to see that no one spends enough time there to be harmed. Title III is, after all, intended to protect residents from air toxics.

The fact that EPA (and other agency) risk-reduction or health-based directives are not well-founded is well known and understood by the professionals involved. The extreme cost of the tiny additions to personal safety associated with some environmental rules has also been well known since before the CAAA were passed in 1990. Yet air toxics policy is one of many similar outcomes of a regulatory process that has been going on for more than twenty years. . . .

A wide array of commentators has noted the problems inherent in the Clean Air Act, its various amendments, and in Title III of the 1990 Amendments. The sources of these problems can be found without resort to asserting stupidity among decision makers at any level, corruption, or lack of dedication among EPA officials. The problems stem from a lack of proper incentives throughout the public sector, including the lack of incentives for voters to become better informed. The move from common law [settling pollution disputes in the courts] to regulation has not been without the introduction of new problems in the case of air toxics. Institutional changes like those suggested can help to bring incentives more in line to help produce a regulatory regime that is less unjust and less wasteful of limited resources.

Chapter 4

Is Recycling an Effective Way to Reduce Pollution?

Chapter Preface

In 1987, a crew navigated a barge full of New York City trash down the East Coast in search of a place to dump the ship's cargo. Unfortunately, there were no takers, and the barge, named *Mobro 4000,* traveled as far as Mexico via the Caribbean Sea before returning to New York City, where its cargo was buried just a few miles away on Long Island. The highly publicized event generated concern that American cities were running out of landfill space in which to bury garbage. However, the *Mobro 4000* was turned away from numerous ports for reasons that had nothing to do with a landfill shortage. As Mary H. Cooper, a researcher with the *CQ Researcher*, reports, "The barge operator had simply set sail without signing any agreements with landfill operators to dump its garbage, and port authorities turned [the barge] away, not because local landfills were full, but because they feared it carried hazardous wastes." Nevertheless, the public's impression that landfills, in addition to polluting the soil and water of adjacent communities, were reaching capacity, encouraged the view that more waste should be recycled.

Beginning in the late 1980s, mayors came to regard recycling as a way to ease fears of a runaway trash problem while reducing the costs of municipal solid-waste disposal. Explains Adam S. Weinberg, author of *Urban Recycling and the Search for Sustainable Community Development*, "Urban mayors were attracted to recycling's promise to reduce the municipal costs of garbage disposal. . . . Instead of paying a 'tipping fee' to place all municipal solid waste into a landfill, cities could collect some of the materials, sort them, and sell them to firms who used them for manufacturing new goods."

Based on these putative benefits, many cities around the country, including New York, Boston, and Seattle, instituted recycling programs. Statistics from the Environmental Protection Agency indicate that the number of curbside recycling programs in the United States grew from 2,700 in 1990 to 8,817 in 1996. The City of New York, which operates the largest mandatory recycling program in the nation, reports that recycling played an important role in the gradual phase-out of the Fresh Kills landfill, which was closed in March 2001.* However, recycling programs in many large cities, including New York, have not reduced the overall costs associated with waste disposal. Cities are sending less trash to landfills, but the market for recycled goods has often been too weak for mayors to recoup the money spent on recycling programs.

To convince municipalities that recycling programs are worth operating, even

*Fresh Kills was temporarily reopened in the fall of 2001 to accommodate debris from the collapsed World Trade Center.

at a loss, environmentalists have maintained that recycling reduces the pollution and waste produced by the manufacture of virgin products, such as newsprint and plastics. They argue that much less waste is generated by extracting usable material from an already manufactured product. But some critics of recycling programs assert that recycling does not offer an environmentally sound alternative to virgin manufacturing. Explains Jerry Taylor, the director of natural resources studies at the Cato Institute, a libertarian think tank, "The actual process of *extracting* usable raw material from a product is an industrial activity every bit as involved as the process of *combining* various raw materials to make a product. . . . [R]ecycling 100 tons of old newsprint generates 40 tons of toxic waste. . . . EPA [Environmental Protection Agency] has reported that 13 of the 50 worst Superfund sites are/were recycling facilites."

Despite charges that recycling neither benefits the environment nor helps local economies, reminders to recycle appear on more and more products, and recycling has become etched in the minds of school children taught the three R's of "reduce, recycle, and reuse." Most major American cities have turned to recycling to control household waste, conserve landfill space, and reduce the environmental impact of virgin manufacturing. Whether recycling is a practice that should be continued and expanded upon is debated in the following chapter.

Recycling Is a Cost-Effective Means to Reduce Pollution

by Richard A. Denison and John F. Ruston

About the authors: *Richard A. Denison is a senior scientist at Environmental Defense, an organization working to solve environmental problems. John F. Ruston is an economist with Environmental Defense.*

Ever since the inception of recycling, opponents have insisted that ordinary citizens would never take the time to sort recyclable items from their trash. But despite such dour predictions, household recycling has flourished. From 1988 to 1996, the number of municipal curbside recycling collection programs climbed from about 1,000 to 8,817, according to *BioCycle* magazine. Such programs now serve 51 percent of the population. Facilities for composting yard trimmings grew from about 700 to 3,260 over the same period. These efforts complement more than 9,000 recycling drop-off centers and tens of thousands of workplace collection programs. According to the Environmental Protection Agency (EPA), the nation recycled or composted 27 percent of its municipal solid waste in 1995, up from 9.6 percent in 1980.

The Recycling Debate

Despite these trends, a number of think tanks, including the Competitive Enterprise Institute and the Cato Institute (both in Washington, D.C.), the Reason Foundation (in Santa Monica, Calif.), and the Waste Policy Center (in Leesburg, Va.), have jumped on the anti-recycling bandwagon. These organizations are funded in part by companies in the packaging, consumer products, and waste-management industries, who fear consumers' scrutiny of the environmental impacts of their products. The anti-recyclers maintain that government bureaucrats have imposed recycling on people against their will—conjuring up an image of Big Brother hiding behind every recycling bin. Yet several consumer researchers, such as the Rowland Company in New York, have found that recycling enjoys strong support because people believe it is good for the environment and conserves resources, not because of government edict.

Alas, the debate over recycling rages on. The most prominent example was a 1996 article that appeared in the *New York Times Magazine*, titled "Recycling Is Garbage," whose author, John Tierney, relied primarily on information supplied by groups ideologically opposed to recycling. Here we address the myths he and other recycling opponents promote.

Landfill Costs

Myth: The modern recycling movement is the product of a false crisis in landfill space created by the media and environmentalists. There is no shortage of places to put our trash.

Fact: Recycling is much more than an alternative to landfills. The so-called landfill crisis of the late 1980s undoubtedly lent some impetus to the recycling movement (although in many cities around the country, recycling gained momentum as an alternative to incineration, not landfills). The issues underlying the landfill crisis, however, were more about cost than space.

Landfill space is a commodity whose price varies from time to time and from place to place. Not surprisingly, prices tend to be highest in areas where population density is high and land is expensive. In the second half of the 1980s, as environmental regulations became more stringent, large numbers of old landfills began to close, and many simply filled up, particularly in the Northeast. New landfills had to meet the tougher standards; as a result, landfill prices in these regions escalated dramatically. In parts of northern New Jersey, for example, towns that shifted their garbage disposal from local dumps to out-of-state landfills found that disposal costs shot from $15–20 per ton of garbage to more than $100 per ton in a single year. Although the number of open landfills in the United States declined dramatically—according to *BioCycle* magazine, from about 8,000 in 1988 to fewer than 3,100 in 1995—huge, regional landfills located in areas where land is cheap ultimately replaced many small, unregulated town dumps. Landfill fees declined somewhat and the predicted crisis was averted. Nonetheless, the high costs of waste disposal in the Northeast and, to a lesser extent, the West Coast, have spurred local interest in recycling: two-thirds of the nation's curbside recycling programs operate in these regions.

> ***"From 1988 to 1996, the number of municipal curbside recycling collection programs climbed from about 1,000 to 8,817."***

But landfills are only part of the picture. The more important goals of recycling are to reduce environmental damage from activities such as strip mining and clearcutting (used to extract virgin raw materials) and to conserve energy, reduce pollution, and minimize solid waste in manufacturing new products. Several . . . major studies have compared the lifecycle environmental impacts of the recycled materials system (collecting and processing recyclable materials and manufacturing them into usable form) with that of the virgin materials system (extracting virgin resources, refining and manufacturing them into usable materials, and disposing of waste through landfills or incineration). Materials included in the studies are those typically collected in curbside programs (newspaper, corrugated cardboard, office paper, magazines, paper packaging, aluminum and steel cans, glass bottles, and certain types of plastic bottles). The studies were conducted by Argonne National Labs, the Department of Energy and Stanford Research Institute, the Sound Resource Management Group, Franklin Associates, Ltd., and the Tellus Institute. All of the studies found that recycling-based systems provide substantial environmental advantages over virgin materials systems: because material collected for recycling has already been refined and processed, it requires less energy, produces fewer common air and water pollutants, and generates substantially less solid waste. In all, these studies confirm what advocates of recycling have long claimed: that recycling is an environmentally beneficial alternative to the extraction and manufacture of virgin materials, not just an alternative to landfills.

"The high costs of waste disposal . . . have spurred local interest in recycling."

Landfills Increase Pollution

Myth: Recycling is not necessary because landfilling trash is environmentally safe.

Fact: Landfills are major sources of air and water pollution, including greenhouse gas emissions.

According to "Recycling Is Garbage," municipal solid waste landfills contain small amounts of hazardous lead and mercury, but studies have found that these poisons stay trapped inside the mass of garbage even in the old unlined dumps that were built before today's stringent regulations. But this statement is simply wrong. In fact, 250 out of 1,204 toxic waste sites on the Environmental Protection Agency's Superfund National Priority List are former municipal solid waste landfills. And a lot more than just lead and mercury goes into—and comes out of—ordinary landfills. The leachate that drains from municipal landfills is remarkably similar to that draining from hazardous waste landfills in both composition and concentration of pollutants. While most modern landfills include systems that collect some or all of this leachate, these systems are absent from older facilities that are still operating. Moreover, even when landfill design prevents leachate from escaping and contaminating groundwater, the collected leachate must be treated and then discharged. This imposes a major expense and burden on already encumbered plants that also treat municipal sewage.

What's more, decomposing paper, yard waste, and other materials in landfills produce a variety of harmful gaseous emissions, including volatile organic chemicals, which add to urban smog, and methane, a greenhouse gas that contributes to global warming. Only a small minority of landfills operating today collect these gases; as of 1995, the EPA estimates, only 17 percent of trash was disposed of in landfills equipped with gas-collection systems. According to a 1996 study by the EPA, landfills give off an estimated 36 percent of all methane emissions in the United States. We estimate that methane emissions from landfills in the United States are 24 percent lower than they would be if recycling were discontinued.

Recycling Costs

Myth: Recycling is not cost effective. It should pay for itself.

Fact: We do not expect landfills or incinerators to pay for themselves, nor should we expect this of recycling. No other form of waste disposal, or even waste collection, pays for itself. Waste management is simply a cost society must bear.

Unlike the alternatives, recycling is much more than just another form of solid waste management. Nonetheless, setting aside the environmental benefits, let's approach the issue as accountants. The real question communities must face is whether adding recycling to a traditional waste-management system will increase the overall cost of the system over the long term. The answer, in large part, depends on the design and maturity of the recycling program and the rate of participation within the community.

Taking a snapshot of recycling costs at a single moment early in the life of community programs is misleading. For one thing, prices of recyclable materials fluctuate, so that an accurate estimate of revenues emerges only over time. For another, costs tend to decline as programs mature and expand. Most early curbside recycling collection programs were inherently inefficient because they duplicated existing trash-collection systems. Often two trucks and crews drove down the same streets every week to collect the same amount of material that one truck used to handle. Many U.S. cities have since made their recycling collection systems more cost-effective by changing truck designs, collection schedules, and truck routes in response to the fact that picking up recyclable refuse and yard trimmings leaves less trash for garbage trucks to collect. For example, Visalia, California, has developed a truck that collects refuse and recyclable materials simultaneously. And Fayetteville, Arkansas, added curbside recycling with no increase in residential bills by cutting back waste collection from twice weekly to once.

"Landfills produce a variety of harmful gaseous emissions . . . which add to urban smog."

Several major cities—Seattle, San Jose, Austin, Cincinnati, Green Bay, and

Portland, Oregon—have calculated that their per-ton recycling costs are lower than per-ton garbage collection and disposal. In part, these results may reflect the overall rate of recycling: a study of recycling costs in 60 randomly selected U.S. cities by the Ecodata consulting firm in Westport, Connecticut, found that in cities with comparatively high levels of recycling, per-ton recycling collection costs were much lower than in cities with low recycling rates. A similar survey of 15 North Carolina cites and counties conducted by the North Carolina Department of Environment, Health, and Natural Resources found that in municipalities with recycling rates greater than 12 percent, the per-ton cost of recycling was lower than that for trash disposal. Higher rates allow cities to use equipment more efficiently and generate greater revenues to offset collection costs. If we factor in increased sales of recyclable materials and reductions in landfill disposal costs, many of these high-recycling cities may break even or make money from recycling, especially in years when prices are high.

Seattle, for example, has achieved a 39 percent recycling/composting collection rate in its residential curbside program and a 44 percent collection rate citywide. Analysis of nine years of detailed data collected by the Seattle Solid Waste Utility shows that, after a two-year startup period, recycling services saved the city's solid waste management program $1.7 to $2.8 million per year. These savings occurred during a period of reduced market prices for recyclable materials; the city's landfill fees, meanwhile, are slightly above the national average. In 1995, when prices for recyclable materials were higher, Seattle's recycling program generated savings of approximately $7 million in a total budget of $29 million for all residential solid waste management services.

> ***"Several major cities . . . have calculated that their per-ton recycling costs are lower than per-ton garbage collection and disposal."***

To reduce the cost of recycling programs, U.S. communities need to boost recycling rates. A study of 500 towns and cities by Skumatz Economic Research Associates in Seattle, Washington, found that the single most powerful tool in boosting recycling is to charge households for the trash they don't recycle. This step raised recycling levels by 8 to 10 percent on average. These kinds of variable-rate programs are now in place in more than 2,800 communities, compared with virtually none a decade ago.

The Market for Recycled Materials

Myth: Recycled materials are worthless; there is no viable market for them.

Fact: While the prices of recycled materials fluctuate over time like those of any other commodity, the volume of major scrap materials sold in domestic and global markets is growing steadily. Moreover, many robust manufacturing industries in the United States already rely on recycled materials. These businesses are an important part of our economy and provide the market foundation for the entire recycling process.

In paper manufacturing, for example, new mills that recycle paper to make corrugated boxes, newsprint, commercial tissue products, and folding cartons generally have lower capital and operating costs than new mills using virgin wood, because the work of separating cellulose fibers from wood has already occurred. Manufacturers of office paper may also face favorable economics when using recycling to expand their mills. Overall, since 1989, the use of recycled fiber by U.S. paper manufacturers has been growing faster than the use of virgin fiber. By 1995, 34 percent of the fiber used by U.S. papermakers was recycled, compared with 23 percent a decade earlier. During the 1990s, U.S. pulp and paper manufacturers began to build or expand more than 50 recycled paper mills, at a projected cost of more than $10 billion.

"The volume of major scrap materials sold in domestic and global markets is growing steadily."

Recycling has long been the lower-cost manufacturing option for aluminum smelters; and it is essential to the scrap-fired steel "mini-mills" that are part of the rebirth of a competitive U.S. steel industry. The plastics industry, however, continues to invest in virgin petrochemical plants rather than recycling infrastructure—one of several reasons why the market for recycled plastics remains limited. Another factor not addressed by the plastics industry is that many consumer products come in different types of plastic that look alike but are more difficult to recycle when mixed together. Makers and users of plastic—unlike those of glass, aluminum, steel, and paper—have yet to work together to design for recyclability.

Trees and Recycling

Myth: Recycling doesn't "save trees" because we are growing at least as many trees as we cut to make paper.

Fact: Growing trees on plantations has contributed to a severe and continuing loss of natural forests.

In the southern United States, for example, where most of the trees used to make paper are grown, the proportion of pine forest in plantations has risen from 2.5 percent in 1950 to more than 40 percent in 1990, with a concomitant loss of natural pine forest. At this rate, the acreage of pine plantations will overtake the area of natural pine forests in the South during this decade, and is projected to approach 70 percent of all pine forests in the country during the next few decades. While pine plantations are excellent for growing wood, they are far less suited than natural forests to providing animal habitat and preserving biodiversity. By extending the overall fiber supply, paper recycling can help reduce the pressure to convert remaining natural forests to tree farms.

Recycling becomes even more important when we view paper consumption and wood-fiber supply from a global perspective. Since 1982, the majority of the growth in worldwide paper production has been supported by recycled fiber, much of it from the United States. According to one projection, demand for paper in Asia, which does not have the extensive wood resources of North America or northern Europe, will grow from 60 million tons in 1990 to 107 million tons in 2000. To forestall intense pressures on forests in areas such as Indonesia and Malaysia, industry analysts say that recycling will have to increase, a prediction that concurs with U.S. Forest Service projections.

Product Prices and Environmental Damage

Myth: Consumers needn't be concerned about recycling when they make purchasing decisions, since stringent U.S. regulations ensure that products' prices incorporate the costs of the environmental harms they may cause. Buying the lowest-priced products, rather than recycling, is the best way to reduce environmental impacts.

Fact: Even the most regulated industries generate a range of environmental damages, or "externalities," that are not reflected in market prices.

When a coastal wetland in the Carolinas is converted to a pine plantation, estuarine fish hatcheries and water quality may decline but the market price of wood will not reflect this hidden cost. Similarly, a can of motor oil does not cost more to a buyer who plans to dispose of it by pouring it into the gutter, potentially contaminating groundwater or surface water, than to a buyer who plans to dispose of it properly. And there is simply no way to assign a meaningful economic value to rare animal or plant species, such as those endangered by clearcutting or strip mining to extract virgin resources. While many products made from recycled materials are competitive in price and function with virgin products, buying the cheapest products available does not provide an environmental substitute for waste reduction and recycling.

Recycling Is Convenient

Myth: Recycling imposes a time-consuming burden on the American public.

Fact: Convenient, well-designed recycling programs allow Americans to take simple actions in their daily lives to reduce the environmental impact of the products they consume.

In a bizarre example of research, the author of "Recycling Is Garbage" asked a college student in New York City to measure the time he spent separating materials for recycling during one week. The total came to eight minutes. The author calculated that participation in recycling cost the student $2,000 per ton of recyclable trash by factoring in janitors wages and the rent for a square foot of kitchen space, as if dropping the newspapers on the way out the door could be equated with going to work as a janitor, or as if New Yorkers had the means to turn small, unused increments of apartment floor space into tradable commodities.

Using this logic, the author might have taken the next step of calculating the economic cost to society when the college student makes his bed and does his dishes every day. The only difference between recycling and other routine housework, like taking out the trash, is that one makes your immediate environment cleaner while the other does the same for the broader environment. Sorting trash does take some extra effort, although most people find it less of a hassle than sorting mail, according to one consumer survey. More important, it provides a simple, inexpensive way for people to reduce the environmental impact of the products they consume.

If we are serious about lowering the costs of recycling, the best approach is to study carefully how different communities improve efficiency and increase participation rates—not to engage in debating-club arguments with little relevance to the real-world problems these communities face. By boosting the efficiency of municipal recycling, establishing clear price incentives where we can, and capitalizing on the full range of environmental and industrial benefits of recycling, we can bring recycling much closer to its full potential.

Recycled Materials Produce Less Pollution

by Allen Hershkowitz

About the author: *Allen Hershkowitz is a leading expert on recycling and a senior scientist at the Natural Resources Defense Council, an environmental conservation group. He is also the founder of the Bronx Community Paper Company, a paper recycling plant in the South Bronx, New York.*

Using recycled materials helps avoid the air and water pollution typically caused by manufacturing plants that rely solely on unprocessed, virgin raw materials. Because using recycled materials reduces the need to extract, process, refine, and transport the timber, crude petroleum, ores, and so on that are necessary for virgin-based paper, plastics, glass, and metals, recycling lessens the toxic air emissions, effluents, and solid wastes that these manufacturing processes create. It is virtually beyond dispute that manufacturing products from recyclables instead of from virgin raw materials—making, for instance, paper out of old newspapers instead of virgin timber—causes less pollution and imposes fewer burdens on the earth's natural habitat and biodiversity.

Antienvironmental theorists dismiss these benefits. The Cato Institute, a conservative research and advocacy group based in Washington, D.C., claims that state and local ordinances that promote recycling "neither conserve scarce resources nor help to protect the environment." According to the Reason Foundation [a conservative think tank], "Recycling itself can cause environmental harm. . . . As a result, the environmental costs of recycling may exceed any possible environmental benefits." Most recently, the benefits recycling provides in avoiding pollution caused at manufacturing plants were dismissed by [columnist] John Tierney [in the 1996 *New York Times* article "Recycling Is Garbage"] as follows: "[T]here are much more direct—and cheaper—ways to reduce pollution. Recycling is a waste of . . . natural resources . . . [and] a messy way to try to help the environment."

Pollution and the Paper Industry

In virtually all cases, recycling helps reduce or eliminate the pollution typically associated with the production and disposal of consumer products. As the following text makes clear, antirecycling interests who argue otherwise are either out of touch with or conveniently ignoring well known and widely documented environmental facts.

Think bundling your newspapers is "messy"? Not when compared with the process of making paper from virgin timber. While modern paper recycling mills produce no hazardous air or water pollution and no hazardous wastes, the virgin pulp and paper industry is one of the world's largest generators of toxic air pollutants, surface water pollution, sludge, and solid wastes. A recent assessment of the virgin-timber based paper-making industry concluded that reducing hazardous discharges at paper mills worldwide to safe levels would cost $27 billion. Indeed, the timber industry has in all likelihood wiped out more habitat and more species per unit of production than has any other industry. Most Americans associate virgin paper mills with both the destruction of resident-species habitat and the contamination of streams and rivers with chlorinated dioxins and other pollutants. But the fact is these mills are also major sources of a wide variety of hazardous air and water pollutants, odors, solid waste, contaminated sludge, and water discoloring agents. Besides their well known, often unbearable emissions of sulfur compounds (causing an odor resembling rotten eggs that wafts offensively for many miles around), pulp and paper mills are classified under U.S. federal law as generators of

> significant quantities of Hazardous Air Pollutants (HAPs) chlorinated and non-chlorinated. Some of these pollutants are considered to be carcinogenic, and all can cause toxic health effects following exposure. Most of the organic HAPs emitted from this industry also are classified as volatile organic compounds which participate in photochemical reactions in the atmosphere to produce ozone, a contributor to photochemical smog.

Moreover, the virgin "pulp and paper industry is the largest industrial process water user in the United States. Approximately 1,551 trillion gallons of wastewater are generated annually by pulp, paper, and paperboard manufacturers." Water pollutants contained in these billions of gallons discharged into streams, rivers, and lakes by virgin paper manufacturers include a wide range of hazardous and conventional pollutants as well as volatile organic compounds, such as chlorinated dioxins and furans, chloroform, absorbable organic halides [AOX], methylene chloride, trichlorophenols, and pentachlorophenols.

"Recycling helps reduce or eliminate the pollution typically associated with the production and disposal of consumer products."

Processing rigid stands of timber into flexible, printable, smooth, glossy (or absorbent) paper requires an intensive chemical and mechanical effort after a tree is harvested. Once roads have been cut into the forest to get to the timber, it is harvested, transported to the mill, stockpiled, debarked, chipped, "cooked" in vats of chemicals, and turned into pulp and bleached mechanically and chemically. Then the pulp must be turned into paper or dried and shipped off to another mill. While paper can be recycled using fewer than a dozen nonhazardous chemicals and bleaching solutions that contain, for example, 99.5 percent water and 0.5 percent hydrogen peroxide (a concentration more diluted than the peroxide in your medicine cabinet), most virgin pulp and paper, by contrast, is made using literally hundreds of highly corrosive and hazardous chemicals, including chlorine. As the Environmental Protection Agency (EPA) has documented, this presents enormous problems in reducing pollution from virgin paper mills because "elimination of dioxin, furan, chlorinated phenolics, and other chlorinated organics . . . [can] not be achieved unless all forms of chlorine-based bleaching are eliminated." This is not expected to happen in the United States for quite some time. In addition, not all of the toxic pollutants discharged in the wastewater produced by virgin pulp and paper mills are currently regulated by the Environmental Protection Agency, including certain congeners of dioxin and furans and a range of chlorinated phenols.

"Most virgin pulp and paper, . . . is made using literally hundreds of highly corrosive and hazardous chemicals."

Plastics

In stark contrast to the manufacture of a ceramic mug, which is environmentally benign, oil refineries and plastics production facilities that process crude petroleum into plastic cups and other consumer goods produce some of the most substantial public health threats—including lethal gases like phosgene—posed by any manufacturing process. In 1994 alone the 1,834 plastics production facilities operating in the United States emitted more than 111 million pounds of toxic air emissions, 507 million pounds of production-related wastes, tens of thousands of pounds of discharges to surface waters, and hundreds of thousands of pounds of other pollutants releases. Moreover, the plastics industry is second only to the chemicals industry in generating toxic releases that damage the ozone layer, emitting more than 12 million pounds of ozone depleting chemicals in 1994. Production of low-density polyethylene (used to make packaging and soda bottles) from virgin resources generates 62 to 92 pounds of organic pollutants per ton of product manufactured. In 1995, this amounted to approximately 500 million pounds of pollutants that needed to be burned, recycled, or discharged. Similar impacts are associated with the production of polystyrene (throwaway cups and foam packaging) and polypropylene (packaging). Finally, plastics production facilities are dangerous to the people who work in them. Between 1983 and 1993, there were forty-six major explosions and fires at oil refineries and chemical plants that produce plastics; these explosions injured 446 workers and killed 89. Virtually all of these impacts are reduced or avoided entirely when recycled plastics substitute for crude oil and gas in the manufacture of consumer packaging and products. For example, recycling plastic containers into resin-pellets or new products involves mostly mechanical processes like shredding and heating and uses very few process chemicals, none of which are hazardous.

The antirecycling chorus says, "Recycling is a messy way to try to help the environment." Remember the Exxon Valdez [oil spill in 1989]? That was rather messy: more than 300,000 birds were killed outright by the spill, as were 3,500 to 5,500 sea otters; the year after the spill there was a 50 percent pink salmon-egg mortality in oiled streams, compared with an 18 percent normal mortality in nonoiled streams. In general, highly productive and biologically rich coastal habitats were severely contaminated. In fact, "thousands of oil spills are reported every year, spilling millions of gallons of oil" (Sankovitch, 1993, p. 3). Much of the oil contained in the Exxon Valdez (and the oil carried by many other tankers that spill each year) was destined for plastics production facilities.

Mining

"Recycling Is Garbage" never specifies exactly why recycling is "a messy way to try to help the environment." Certainly no discussion about making an industrial-strength mess is complete without a reference to minerals and metals mining. Metals recycling—of steel cans, aluminum, iron, and so on—is one of American industry's success stories, helping to reduce costly imports and reducing the need for virgin ore mining. It does seem hard to believe that recycling can be nearly as messy as mining virgin resources. Unlike the mining of virgin ores, reclaiming "urban ores," the metals found in the U.S. municipal waste stream, does not involve cutting open vast tracks of land, constructing huge mechanical land-movers to churn up the earth, contaminating the air, nearby rivers, lakes, and streams. Quite the contrary, recycling works to avoid some of the disastrous yet typical mining impacts described by the Mineral Policy Center (MPC), a bipartisan research group. The MPC estimates that more than 550,000 hardrock abandoned mine reclamation (HAMR) sites exist nationwide. Abandoned mines, visual eyesores that leach pollutants, are found in every state of the union.

> ***"Metals recycling . . . is one of American industry's success stories."***

To name but a few of the sites where mines have created serious problems:

- *Alaska*: A long-defunct antimony mine is located in Denali National Park, home of Mount McKinley. The tailings pile and tunnels from this mine have begun to contaminate nearby scenic Slate Creek.

• *Arkansas*: Bauxite pits scattered throughout the state pose not only environmental but also safety and environmental threats. Between 1986 and 1988, six people drowned in water-filled pits, known locally as "blue holes." Miles of streams have also lost all life because of bauxite contamination, including the once-thriving Blue Branch Creek.

• *Idaho*: The Cinnabar mercury mine, in central Idaho, has been discharging mercury into a nearby river. Local children wandering onto the site are exposed to mercury and other hazardous substances.

Recycling metals and minerals—tin cans (tin, iron), batteries, (cadmium, zinc, mercury, lead, manganese) and aluminum (bauxite)—as well as motor oil, paper, and plastics, as minor local ordinances encourage millions of Americans to do, supplements the supply of raw materials and thus reduces the need for mining, timber harvests, and petroleum refining.

Pollution and Paper Recycling Mills

The most controversial charge leveled by the antirecycling crowd is that recycling really doesn't benefit the environment; that, to the contrary, it produces its own significant pollution. For example, the Cato Institute has claimed that "de-inking 100 tons of old newspaper for subsequent reuse generates 40 tons of toxic waste." According to the Reason Foundation, "The environmental costs of recycling may exceed any possible environmental benefits." Here's how John Tierney put it:

> [R]ecycling operations create pollution in areas where more people are affected: fumes and noise from collection trucks, solid waste and sludge from the mills that remove ink and turn the paper into pulp. Recycling newsprint actually creates more water pollution than making new paper from virgin sources. . . . For each ton of newsprint that's produced, an extra 5,000 gallons of waste water are discharged.

Actually, less water pollution is produced per ton by paper recycling mills than by virgin paper mills. The recycled sector of the global paper industry, being more recently developed, is in fact the industry's most modern, efficient, and least polluting sector. The charge made by the Cato Institute that newsprint recycling mills produce forty tons of "toxic" waste for every hundred tons of paper recycled is, simply, absurd. Far from producing more hazardous pollution than virgin mills, modern paper recycling mills produce no hazardous air or water pollution or hazardous wastes. Even the most cursory review of engineering designs for newsprint recycling mills reveals that the product yield per ton of recovered paper used by the mill is in the range of 90 percent—in other words, only about ten tons end up as waste per hundred tons of manufactured paper. None of this waste is "toxic." For virgin mills, the ratio is virtually the opposite: 75 percent of the harvested tree does not wind up as paper product. Where would the "toxic" residues allegedly produced by paper recycling mills come from? Less than one percent of the waste from a recycled paper mill is from ink, which is today more properly described as benign vegetable dye or carbon coated with plastic polymers; the remaining waste is water (90 percent) and short paper fibers (about 10 percent). No recycled paper mill could operate successfully, financially or otherwise, with a 60 percent product yield, as the Cato Institute has claimed, especially if the by-product generated was "toxic waste," which requires special, extremely costly handling, treatment, and disposal according to U.S. federal hazardous waste laws.

Contrary to the claim that the recycling of newspaper produces an extra "5,000 gallons of waste-water" per ton of newsprint manufactured, new paper mills that recycle 100 percent newsprint don't even consume or discharge a total of 5,000 gallons of water per ton of manufactured product. By contrast, the virgin "pulp and paper industry is the largest industrial process water user in the United States." Virgin newsprint mills, at best, use about the same amount of water as recycled newsprint mills, but in fact most virgin newsprint mills use more water per ton of manufactured product, sometimes twice as much. For example: A 100 percent recycled-newsprint mill in Aylesford, England, uses and discharges less than 4,000 gallons of water per ton of manufactured product. A 100 percent recycled-newsprint mill under development in New York City's South Bronx will consume about 3,800 gallons of water per ton of manufactured product, and more than 80 percent of that water will come from a sewage treatment plant as recovered and cleaned effluent. Another mill under construction on Staten Island in New York City, which will recycle newspapers and other types of wastepaper, will also use less than 4,000 gallons of water for every ton of manufactured product. By contrast, virgin newsprint mills that are ten years old or older use approximately 10,000 gallons of water per ton of manufactured newsprint, while the few modern mills that have been built in the past eight to ten years use 4,000 to 5,000 gallons of water per ton of manufactured newsprint. And most newsprint purchased in the United States is produced at older Canadian mills. Overall, the recycled paper industry is evolving as the most modern, efficient, and least polluting sector in the paper manufacturing industry.

"Less water pollution is produced per ton by paper recycling mills than by virgin paper mills."

How did the author of "Recycling Is Garbage" come up with the claim that more water pollution is caused by recycling newsprint? He has indicated that he took the data on water pollution from a chart found in a study on recycled paper produced by the Paper Task Force, which was comprised of the Environmental Defense Fund, Time Inc., Johnson & Johnson, and other paper purchasers. But the chart from which "Recycling Is Garbage" excerpted data was in a report that concluded that more recycled paper should be purchased. According to the report's conclusion: "This analysis shows clear and substantial environmental advantages from recycling all of the grades of paper we examined." The bar chart that "Recycling Is Garbage" used for water-pollution data consists of fourteen graphical comparisons regarding energy usage, atmospheric emissions, solid wastes, and waterborne wastes. Although eleven of the fourteen comparisons show recycled newsprint production having lower environmental impacts than comparable virgin production, "Recycling Is Garbage" leads readers to believe just the opposite. Checking with a paper company that actually operates a newsprint manufacturing facility would have been helpful (as was done for this viewpoint), especially given the fact that the Paper Task Force deliberately avoided including manufacturers of newsprint or any other paper grade in the report "Recycling Is Garbage" refers to.

Transporting Recyclables

It is also wrong to argue that pollution from recycling vehicles in urban areas is greater than pollution from trucks and infrastructure dedicated to garbage collection and disposal. At worst, recycling trucks and infrastructure may produce the same amount of pollution. But, as explained below, it is more likely that diverting waste to recycling processing facilities tends to cause less pollution than garbage collection does.

Any added pollution that might be generated by a recycling program would result from additional trucks dedicated to collection. However, because in most cities "the fleet of trucks to collect recyclables is substantially smaller and less costly than the [typical] waste [collection] fleet," the types of vehicles used to collect recyclables typically generate less pollution than standard waste collection vehicles. Moreover, collecting recyclables has been shown to be a faster operation than collecting garbage. "It takes less time at the curb to pick up a load of recyclables, which is typically under 10-15 pounds, than solid waste, which can be as much as 50 pounds or more." This means a truck collecting recyclables, on average, idles for a shorter time than does a garbage truck, thus emitting less pollutants. Finally, by diverting waste into smaller, less polluting recycling vehicles, real-world experience confirms that there can be a reduction in the number of larger trucks and, indeed, entire truck routes dedicated to garbage collection. As the recycling managers in Madison, Wisconsin have found,

> recycling makes it possible to reduce the number of packers needed to collect solid waste. When 20 percent to 30 percent of the weight (and even more of the volume) is diverted from the refuse trucks to the recycling collection trucks, fewer compactors [for waste] will be necessary. . . . [Moreover, with] less material going into the refuse trucks [per route], those vehicles can usually stay on their routes all day without having to make the typical mid-day trip off-route to unload [and] the time at each stop is reduced because there is less to collect. . . . [T]hat reduces the number of trips off-route to unload from two to one each day [and] reduces the number of routes [needed entirely].

The higher the recycling rate, the fewer trucks needed for garbage collection. Thus, the strategy to reduce any additional pollution that duplicate trucks might cause is to increase recycling. In that way trucks dedicated to garbage collection can be retired.

Moreover, because landfills are usually located far from the concentrated population centers that generate municipal wastes, transporting garbage to landfills typically requires more vehicle miles to be traveled than does the deposit of recyclables at processing plants, which are usually numerous and more frequently located within city limits. In New York City, more than two dozen recycling processing plants conveniently located in every borough help reduce the vehicle miles traveled to deposit recyclables. By contrast, the nonrecyclable portion of the city's waste is deposited at any one of only eight marine transfer stations by the city's fleet of 2,500 collection vehicles, or directly at the logistically remote Fresh Kills landfill on Staten Island. Were New York City to adopt a policy of long-hauling waste for export, as dozens of other cities do, even more vehicle miles would be traveled, increasing mobile-source pollution.

Efficient Recycling Programs Can Reduce Waste

by Brenda Platt and Neil Seldman

About the authors: *Brenda Platt is the director of materials recovery at ILSR. Neil Seldman is president of the Institute for Local Self-Reliance (ILSR), a research organization that promotes environmentally safe economic development.*

In the last decade and a half, waste prevention, reuse and recycling have made tremendous gains. The national municipal recycling rate has reached 28%, while many communities are surpassing 50% diversion from landfills and incinerators, and doing so cost-effectively. More than 9,300 communities had curbside recycling programs in 1998, up from 2,700 at the beginning of the decade. Reduction of private sector and industrial process wastes has similarly increased, with some businesses approaching 90% and higher waste reduction levels.

Recycling's Benefits

The benefits of waste reduction are more far reaching than previously thought. Recycling reduces costs, creates jobs and businesses, and improves the environment and public health in myriad ways. When a pound of municipal material is recycled, industry avoids wasting many more pounds of mining and manufacturing wastes caused by extracting and processing virgin materials into finished goods. Using recycled materials to make new products saves energy and other resources, reduces greenhouse gases and industrial pollution, and stems deforestation and damage to fragile ecosystems.

Waste reduction also reduces the negative effects of landfilling and burning materials. For landfills, these effects include groundwater pollution, release of global warming gases, and monitoring and remediation costs that will likely span centuries. Incinerators may even be worse, as pollution is borne directly to the air as well as to the land as ash; and energy wasted by not recycling is greater than the amount of energy produced via burning.

Despite these benefits, unsustainable patterns of wasting and consumption hinder further progress in recycling. Recent trends indicate wasting is on the rise and is outpacing the rise in recycling:

- After several years of leveling off and then dropping, municipal wasting increased again in 1997, both in absolute tons and on a per capita basis. Materials landfilled and incinerated rose 4.4 million tons in 1997 (the latest year for which data are available) as compared to 1996.
- Municipal recycling rates since 1994 have increased only slightly, after rapidly increasing in the late 1980s and early 1990s.
- The portion of plastic, aluminum, and glass containers landfilled and burned is rising. In 1998, 75% of plastic PET (no. 1) containers were wasted, up from 60% in 1995. The wasting rate for aluminum cans has climbed from a low of 36% in 1992 to 44% in 1998.
- Manufacturers are producing more products and packaging that are hard to recycle or lack recycled-content. From 1990 to 1997, plastic packaging grew five times faster by weight than plastic recovered for recycling.
- The waste hauling industry continues to consolidate, leading to less recycling. Big hauling companies that are vertically integrated with wasting facilities make more money by landfilling than recycling.

"The benefits of waste reduction are more far reaching than previously thought."

- Some states are considering rescinding recycling goals and policies. A few cities have opted to cut back their recycling budgets. Some industries—particularly the plastic industry—have not followed through on commitments to utilize more recycled material.

Several factors contribute to the increase in wasting. For one, manufacturers and sellers of products and packaging usually have no responsibility for handling materials once discarded. Secondly, recycling competes with raw materials processing and wasting industries on an uneven economic playing field:

- Prices of virgin materials and products (which compete with recovered materials) exclude billions of dollars in taxpayer subsidies, and the true costs that resource extraction and manufacturing impose on the environment and public health.
- Prices for waste disposal (which competes with reuse and recycling for the supply of discarded materials) do not reflect the cost of perpetual landfill maintenance, among other externalities.
- The economic development benefits of recycling are often overlooked (recycling creates at least ten times more jobs than landfills).

Zero Waste and a Sustainable Future

We need a new paradigm for managing resources sustainably. Zero waste is a design principle for a society that makes products with a minimum investment of natural resources and energy, and in which the end-of-life options for those products are limited to reuse, recycle, repair, and compost. Zero waste implies that the goal of public policy should be to eliminate waste rather than manage it in waste facilities.

Fortunately, technological developments, citizen activism, and public policies in the last 15 years have laid the groundwork for a zero waste and sustainable future. Container deposit laws, curbside collection, recycling requirements, landfill disposal bans, and creative funding mechanisms have increased the supply of recyclable materials. States with minimum recycled-content legislation, buy-recycled programs, and creative funding mechanisms have also begun to spur demand for discarded materials and link recycling with local economic development. Much more remains to be done to reduce waste and increase reuse and recycling.

Recycling is fundamentally cheaper than wasting when all costs are considered. In addition to providing net pollution prevention benefits, recycling adds value and jobs to local and regional economies.

Recycling is a win-win proposition when we account for (1) upstream subsidies for virgin resource extraction industries, (2) downstream subsidies for landfills and incinerators, (3) the true long-term societal and environmental costs of resource extractive and wasting facilities, and (4) the local economic development benefits of reuse and recycling.

"Zero waste implies that the goal of public policy should be to eliminate waste rather than manage it in waste facilities."

Some state and local governments are improving accounting techniques for evaluating discard management options. Florida, North Carolina, Indiana, and Georgia are some states that promote some level of "full-cost accounting" (although these methods do not incorporate major categories such as subsidies and environment externalities). Local governments already using "full-cost accounting" techniques include Plano, Texas; Sacramento, California; and Seattle, Washington. Still, these techniques need refinement to truly account for appropriate remediation, contingent, environmental, and social costs.

Reducing Costs on an Unlevel Field

But even with an unlevel playing field, many businesses and communities that prevent waste and recycle have reduced their costs. The U.S. Environmental Protection Agency (EPA) reports that in 1997, its WasteWise partners—businesses and institutions that commit to reducing their waste—saved an estimated $218 million in avoided disposal fees alone through recycling efforts. Avoided paper purchasing costs for all reporting partners in 1997 may have been as high as $60 million.

Local governments can also save. A . . . U.S. EPA study of 14 communities recovering between 44% and 65% of their residential waste, found that 13 of these had cost-effective programs. Other research shows that costs for recycling decrease as recovery levels increase. One factor for this is the costs for processing recyclables and yard debris are often much less than landfill or incinerator disposal tip fees.

Restructuring waste management systems can pay off handsomely. For example, Madison, Wisconsin, reduced trash routes by 32% and switched to smaller trash trucks, after introducing its multi-material curbside recycling and yard debris collection programs. These trucks cost less and have lower repair costs than the trucks the city needed to collect all discarded materials as trash. The overall collection cost went down in Madison compared to the cost of operating a single fleet to pick up unseparated waste. Falls Church, Virginia, reduced trash collection frequency from twice to once a week, one year after implementing a multi-material curbside recycling program. As a result, the city raised its material recovery rate from 39% to 65%, cut trash collection costs by more than half, and reduced annual per household waste management costs by more than one third.

"The costs for processing recyclables and yard debris are often much less than landfill or incinerator disposal tip fees."

In some communities recycling is viewed as an expensive burden. But often that is because these communities are recycling at low rates and are treating recycling as an add-on to their traditional trash system rather than a replacement for it. When communities reach high waste reduction levels, recycling becomes more cost-effective. Communities that maximize recycling save money by redesigning their collection schedules and/or trucks. Staff once devoted to trash collection now collect recyclables or yard trimmings. As communities attain ever higher recovery levels, planners and public works administrators are beginning to realize that recycling and composting can be the primary strategy for handling discards, rather than a supplement to the conventional system. The economics of recycling improves when, instead of adding costs of recycling onto the costs of conventional collection and waste disposal, recycling becomes the heart of the system.

Recycling Means Business

Recycling is an economic development tool as well as an environmental tool. Reuse, recycling, and waste reduction offer direct development opportunities for communities. When collected with skill and care, and upgraded with quality in mind, discarded materials are a local resource that can contribute to local revenue, job creation, business expansion, and the local economic base.

On a per-ton basis, sorting and processing recyclables alone sustain 10 times more jobs than landfilling or incineration. However, making new products from the old offers the largest economic pay-off in the recycling loop. New recycling-based manufacturers employ even more people and at higher wages than does sorting recyclables. Some recycling-based paper mills and plastic product manufacturers, for instance, employ on a per-ton basis 60 times more workers than do landfills.

Value is added to discard materials as a result of cleaning, sorting, and baling. Manufacturing with locally collected discards adds even more value by producing finished goods. For example, old newspapers may sell for $30 per ton, but new newsprint sells for $600 per ton. Each recycling step a community takes means more jobs, more business expenditures on supplies and services, and more money circulating in the local economy through spending and tax payments.

Recycling has had a major impact on job creation in local and state economies:

• In North Carolina, recycling industries employ over 8,700 people. The job gains in recycling in this state far outnumber the jobs lost in other industries. For every 100 recycling jobs created, just 10 jobs were lost in the waste hauling and disposal industry, and 3 jobs were lost in the timber industry.

• A survey of ten northeastern states found that they employ 103,413 people in recycling.

• A 1992 survey in Washington found that this state had created 2,050 recycling-based jobs since 1989.

• Massachusetts employs more than 9,000 people in more than 200 recycling enterprises. About half of these jobs are in the recycling-based manufacturing sector. These businesses represent more than half a billion dollars in value added to the state's economy.

"Recycling is an economic development tool as well as an environmental tool."

• In California, meeting the state's 50% recycling goal is expected to create about 45,000 recycling jobs by the year 2000, over 20,000 of which are slated to be in the manufacturing sector.

Regional studies of employment and the remanufacturing industry indicate that recycling activities employ more than 2.5% of manufacturing workers. Extrapolating these findings to the entire nation, recycling and remanufacturing activities could account for approximately 1 million manufacturing jobs and more than $100 billion in revenue.

Reuse Is Best Use

Product reuse is even more job-intensive than recycling. It is a knowledge-based industry, with a premium placed on accurate sorting and pricing, and good inventory management. One reuse company is Urban Ore in Berkeley, building materials to books and art. Materials are sorted and cleaned, and sometimes repaired. For the most part, what does not sell becomes scrap. Urban Ore calculates value-added monthly, which ranges from 30% to 60%. This reflects the large contribution its staff and handling system make to its monthly income. As in recycling, Urban Ore is the first link in a value-added chain that involves and employs hundreds of remodeling and landscape contractors, artists, inventors, builders, collectors, property managers, homeowners, and second-hand dealers.

The reuse industry competes with mass-marketed commodities such as diapers, tires, and plastic, glass, and metal drink containers. Each year Americans spend billions of dollars on these new products. Some of this money remains in communities where the products are purchased, but most leaves the community for the home offices of the corporations. A handful of companies dominate the markets for soft drinks, disposable diapers, and new tires.

By contrast, reuse industry alternatives—refillable bottle washing plants, cloth diaper services, tire retreading enterprises—create wealth and jobs for local communities. Such reuse companies tend to be small and locally owned and operated, providing local jobs and increased capital retention. Reuse is thus a tool for miniaturizing global and national economies, making them more sustainable.

There are 1,700 tire retreading operations in North America. About 95% of these are small businesses. Reusable diaper services employ 10,000 to 12,500 people. Each business employs 5 to 50 workers. A complete switch to diaper services would generate 72,000 jobs nationwide in this service industry alone.

Other reuse efforts can have similar impacts. For instance, if building deconstruction were fully integrated into the demolition industry, at least 100,000 jobs could be created in this sector.

Recycling Programs Are Unprofitable and Unnecessary

by Christopher Douglass

About the author: *At the time this viewpoint was written, Christopher Douglass was the John M. Olin Fellow in Public Policy at the Center for the Study of American Business. He went on to enroll in the doctoral program in government and foreign affairs at the University of Virginia.*

Recycling has become a major social institution in America. Its symbol—three chasing arrows—is ubiquitous, seemingly adorning every consumer product. Drop-off areas and recycling bins dot the suburban landscape. Elementary school students learn a new "Three R's"—Reduce, Reuse, Recycle.

The Growth of Recycling in America

Almost 200 million Americans have access to community-run recycling programs, according to a survey by the American Forest and Paper Association. There are nearly 9,000 curbside recycling programs serving 52 percent of the American people. According to one survey, the recycling rate for the nation's municipal solid waste grew from 10 percent in 1985 to 30 percent in 1998.

Popular among consumers and environmental groups alike, recycling has become America's environmental good deed for the day. Allen Hershkowitz of the Natural Resources Defense Council says, "Recycling is probably the single most successful environmental policy out there. Most people in the world today know about reduce, reuse, recycle. It is very widely practiced. More people participate voluntarily in recycling than voted in the last four presidential elections."

Though it has recently escalated in popularity, recycling, as a practice, is not new. Merchants have collected cardboard boxes, baled, and then recycled them for decades. An industry for salvaging and reprocessing automobiles and large appliances has operated for half a century. It was not until recently, however,

Excerpted from "Government's Hand in the Recycling Market: A New Decade," by Christopher Douglass, *Policy Study Number 148*, September 1998.

that government at all levels elevated recycling to a national priority.

In the late 1980s, dramatic predictions of landfill closings and a vagabond trash barge loaded with New York City trash created a crisis mentality in America. A 1988 Environmental Protection Agency (EPA) report to Congress projected that one-third of all landfills in the United States would close by 1994 and that by 2008 nearly 80 percent of landfills would be shut down. EPA Assistant Administrator J. Winston Porter claimed at the time, "We have a real [landfill] capacity crunch coming up." Senator Max Baucus (D-Montana), then chairman of the Senate Subcommittee on Environmental Protection, said, "We are overwhelming ourselves with garbage, and we are running out of safe and secure places in which to place it."

State and local governments responded to the resulting public concern by turning to recycling. States established recycling goals and requirements, grants for recycling programs, and mandates for recycled products in government purchases. Forty-four states established recycling goals in the late 1980s. Illinois, for instance, passed a law mandating county recycling plans and specifically cited dwindling landfill space as its justification. Local communities expanded their trash collection services to include curbside recycling programs and recycling drop-off sites.

Beyond solving a crisis, further justifications for these efforts included the preservation of resources and, in the end, creation of a garbage disposal method that would avoid the pollution of landfilling or incineration. Plus, cities were told that after markets for recyclables matured, local governments would make money by selling recovered household waste. Recycling promised to be an economical and environmentally beneficial answer to the nation's garbage woes. The timing was propitious for new infrastructure investment, as 1989 saw most Americans choose the environment as their top priority for more government spending, ahead of even crime and health care, according to a National Opinion Research Center poll.

Although the recycling movement was spectacularly successful at motivating action, its foundational notions were, in large part, misconceptions. There was, in fact, no landfill crisis in the late 1980s, and recycling is not always the most environmentally responsible way to manage household waste.

"Recycling has become America's environmental good deed for the day."

The trash barge's well-publicized failure to find harbor [in 1987] was not the result of a lack of landfill space. An entrepreneur had filled the Mobro 4000 with New York City trash and planned to dump it in the South, where landfill fees were lower due to the abundance of landfill space. Because he had not found a dumpsite before setting sail, a landfill in North Carolina suspected toxic waste might be mixed in with the garbage and rejected the Mobro's load. After heavy media coverage of this refusal, no other community would accept the trash. Following two months of sailing, the Mobro 4000 returned to New York, where its load was incinerated.

The nation had lost landfills, but it had lost little landfill space. New federal regulations to limit water pollution and gas emissions raised the average cost per metric ton of operating landfills from $9 to $20. Small, mainly publicly-owned landfills opted to close rather than incur the high cost of meeting the new standards. Although high numbers of landfills closed, the new landfills opening in the 1990s were much larger in order to offset the high fixed costs of the new rules.

"The recycling movement['s] . . . foundational notions were, in large part, misconceptions."

During the supposed crisis, sites were abundant for new landfills, even in the densely populated Northeast. A 1989 survey of the eastern half of the state of New York performed by the state government found more than 200 square miles of geologically safe land. This amount represents less than 0.4 percent of the state's area, yet it could hold all of New York's trash for the next two centuries.

Today, landfill capacity is more than adequate. When asked in a 1997 survey whether there was a disposal capacity problem in their state, 45 out of 46 states responded that there was no capacity problem, with only Hawaii reporting inadequate landfill space. Twenty-nine of 37 states responding to a 1998 survey said they have more than ten years of capacity available, with the state of Wyoming claiming 100-plus years of capacity.

Environmental Benefits Questioned

In addition, some . . . dispute that recycling is always environmentally beneficial. Matthew Leach, Ausilio Bauen, and Nigel J.D. Lucas point out in the *Journal of Environmental Planning and Management* that the more one values clean air, the more recycling should rankle, due to the long distances recyclables must usually be transported by pollution-spewing, energy-consuming vehicles. Additionally, Leach, et al., find that every ton of paper incinerated rather than recycled in Great Britain prevents 300 kilograms of carbon dioxide emissions.

Others have pointed out that de-inking newsprint or office paper can create hazardous sludge. Ghislain Bolduc, manager of a paper recycling plant, told *Chemical Marketing Reporter* in 1994, "Newspaper consists of about 2 percent ink, but the concentrate of sludge we acquire is toxic and it has to be dealt with. For every ton of mixed office waste paper that we de-ink, we create roughly one ton of sludge."

Recycling technologies such as de-inking are becoming progressively cleaner, but transporting, sorting, and processing recyclables cannot be assumed to have zero environmental costs.

Growing public recognition of these fundamental misconceptions has borne little effect on government action in recent years. Government at all levels continues to sponsor and support recycling through a myriad of pick-up programs, educational initiatives, business subsidies, content mandates, and purchasing preferences. . . .

Curbside Recycling Programs: Operating at a Loss

Municipally-run curbside recycling programs ballooned from just over 1,000 in 1989 to 9,000 in 1998. For nearly all local governments, extending this service to constituents has been expensive. The cost of collecting and sorting recyclables has exceeded their market prices in most parts of America, forcing most recycling programs to operate at a deficit.

Franklin Associates, the consulting firm which compiles EPA's annual solid waste report, estimates that the cost of collecting recyclables is about $139 per ton for programs that recycle old newsprint and magazines, steel and aluminum cans, plastic packaging, and glass containers. The cost of sorting these recyclables averages $86 per ton, and the benefits from avoiding landfilling fees is typically $27 per ton, for a net cost of $198 per ton. The majority of recyclables collected yield less than $198 per ton at 1998 prices. Aluminum cans have been profitable for some time and HDPE and PET, the two most common plastics for soda, milk, and water containers, now cover their costs, PET only recently. Although these prices and costs vary across the country, this table provides a rough idea of the disparity between most recycled material prices and the costs of municipal curbside recycling.

The bottom line for many municipal recycling programs shows a net loss. The costs incurred by collecting recyclables from the curb, transporting them to a materials recovery facility, and sorting them there cannot be recovered in the current recyclables market. Calculations by Franklin Associates in 1997 put the average costs for municipal waste programs with curbside recycling at $144 per household compared to costs of $119 per household for cities without recycling programs. Table 1 shows the per household cost of recycling programs in eight localities of varying size around the nation, ranging from an estimated $13 per household in Aurora, Illinois to $36 per household in Portland, Oregon.

"During the supposed [landfill] crisis, sites were abundant for new landfills."

Forced Cutbacks

In the Baltimore metropolitan area, the city and its six surrounding counties have recycled a total of 2.5 million tons since 1991 through curbside recycling programs and other residential programs, at a total cost of $250 million. All of those recyclables collected could have been buried in 27 acres in the nearby King George County Landfill at a total cost of $83 million, according to *The*

Baltimore Sun, 33 percent of the cost of recycling. Similarly, in 1997, the small town of Sanford, Maine, paid a company $90,990 to collect recyclables that could have been landfilled for $13,365.

Although state governments have alleviated some of this cost burden through grants, many localities have been forced to cut back their recycling programs or discontinue them altogether. The county waste agency for Des Moines, Iowa, is looking for ways to continue its recycling program despite incurring losses between $500,000 and $1 million in 1997. Dumping fees at the county landfill had paid for the program until haulers began going to less expensive sites. To avoid these losses to the county landfill, the waste agency cut the fees and asked cities in the spring of 1998 to begin paying the $25.20 annual cost per household for recycling services.

Table 1. Costs of Selected Municipal Recycling Programs

Locality	Annual Program Cost	Number of Households Served	Annual Cost Per Household
Portland, OR	$4,680,000	130,000	$36.00
Santa Fe, NM	637,000	22,000	29.00
Chillicothe, OH	220,000	8,000	28.00
Malibu, CA	96,000*	4,000	24.00
Prince George's County, MD	2,900,000	150,000	19.00
Nashville, TN	1,200,000	70,000	17.00
Cambridge, MA	657,000	41,000	16.00
Aurora, IL	428,000**	33,000	13.00

Notes: * Haulers charge $2.00 per month per household for recycling services.
** Estimate calculated from bid price of $0.19 per bag, 2.25 million bags.

The city council of 24,000-resident Chillicothe, Ohio, dropped its $220,000 recycling program in January 1998, citing fewer state and federal grants, low recyclables prices, and more important city needs, such as a new aerial ladder truck for their fire department. Chillicothe councilman Paul Thurman said, "To me, it's [the recycling program] just a waste of tax money."

In the nine states where curbside recycling programs are mandated, localities have felt the pressure to meet state-imposed goals. Officials in Prince George's County in Maryland, for instance, say that it has been state pressure more than constituent desire that has driven its program to continue operating. Chelo Cole, head of the county waste reduction program, says, "We have to answer to the state. Right now the pressure from the state is to achieve a 20 percent waste reduction." In Florida the law allows the state government to withhold permits and deny grants when counties fail to meet their goals, although such sanctions have not yet been applied.

Policing Residents

Some localities have responded by monitoring and policing their residents to ensure individuals contribute to their area's recycling rate. Mary Andrews, recycling coordinator for the town of York, Maine, says, "We feel like some residential people are not recycling like they should, so we're toying with the idea of using clear bags for trash, so if the hauler sees milk cartons and other recyclable things in there, he wouldn't pick up their trash."

Officials in Hampton, Virginia, are using a radiomagnetic scanner system to keep track of which homes are putting out their recycling bins. An antenna mounted on recycling trucks uses radio waves to register small button-like tags on the recycling bins that it passes by. If a household's recycling bin is placed on the curb at least four times during an eight or nine week period, it is charged three dollars less than households that have not put out their bins. City officials claim that recycling participation has jumped to 70 percent since the scanner system was implemented.

"Transporting, sorting, and processing recyclables cannot be assumed to have zero environmental costs."

More common than these methods are education and publicity. An article written for cities trying to begin recycling programs notes, "Public support is imperative before program implementation. Many communities have found it pays to hold public meetings, mail informational pamphlets, use volunteers to distribute literature and answer residents' questions, erect banners and signs promoting the program, and work with the media in getting the word out. Without a good awareness campaign, your program may have difficulty achieving a high participation rate." Maine's state recycling manager says, "It's got to be made part of the culture of the community."

Such social-awareness initiatives have been instrumental in increasing recycling collection rates. These programs have remained unprofitable, however, because there is little demand for recyclable materials at prices that would cover collection and processing costs. Governmental activity to increase demand for these materials has been steady since the 1980s and has risen slightly in recent years. . . .

Reevaluating Support for Recycling

A reassessment of recycling policy is necessary in view of a decade of experience. Early justifications for recycling were based on several misconceptions, and recycling programs have not been the profit centers municipalities thought they would become. As America enters its second decade of widespread recycling, rates are slowing their increase, and profitable markets for recyclables have largely failed to develop.

A major difficulty that Americans should recognize is that there is no single, simple waste management method that applies to all communities. In constructing an economic model of solid waste management, economist Jannett Highfill and mathematician Michael McAsey came to the conclusion, "Not only is no single recycling plan going to be optimal for all cities, even the barest outlines of the waste management program differs between communities."

Local governments must return to a fundamental question: What environmental and economic benefits are produced by recycling and at what cost? Cataloging environmental benefits, landfill costs avoided, prices being paid for recyclables, and a number of other factors can be difficult, but is essential to determining whether recycling makes sense for any given community. As politicians and waste professionals consider their options in light of a decade of experience, economic calculations of costs and benefits must go hand in hand with the non-monetary benefits often cited as justification for continuing recycling programs.

Only a handful of commodities have shown themselves to earn more than the costs that cities on average incur in collecting and sorting recyclables. Despite its significant buying power, government efforts to prop up recycling markets through increasing demand for items such as recycled office paper have been insufficient to make widespread recycling profitable.

In recent years, many have come to recognize a disparity between the role recycling plays at the practical, day-to-day level and what is being discussed by environmentalists and policymakers as serious environmental problems. Frank Ackerman, author of *Why Do We Recycle?,* points out,

> Recycling is a favorite topic of books full of "household hints to help save the planet;" nothing, it seems, is better suited for do-it-yourself environmental improvement than household waste. But books analyzing the fate of the earth and the state of the environmental movement have almost nothing to say about recycling and solid waste.

In light of the misperceptions that fueled recycling's rise and government's modest progress in developing markets for recovered materials, America should reevaluate its support for recycling. In many cases, funds now spent on unprofitable recycling programs would be better applied to more necessary services. At a minimum, state and local officials would do well to better inform their citizens of the costs and benefits of recycling.

Mandatory Recycling Can Waste Resources

by Alexander Volokh and Lynn Scarlett

About the authors: *Alexander Volokh is an environmental researcher at the Reason Public Policy Institute (RPPI), a conservative think tank. Lynn Scarlett is the executive director of RPPI and the author of several papers on environmental issues.*

"What a Waste. Recycling squanders money and good will—and doesn't do much for the environment, either." This was the cover blurb for John Tierney's story in *The New York Times Magazine* picturesquely titled "Recycling Is Garbage." Lest the reader miss the point, the subhead in the article explained: "Rinsing out tuna cans and tying up newspapers may make you feel virtuous, but recycling could be America's most wasteful activity." This was, of course, anathema to most mainstream environmental organizations, who have spent decades trying to convince Americans to "reduce, reuse, recycle." These "three Rs" of the environmental movement are sometimes called the "solid waste hierarchy," with source reduction—using less stuff—being better than directly reusing old products, which in turn is better than recycling. Recycling, of course, is better than all the other options—incineration, say, or, God forbid, landfilling. Environmental groups responded to Tierney's article with a flurry of studies of their own, essentially restating their longtime position that recycling is "perhaps the most basic . . . of all environmental policies."

A Complicated Picture

Unfortunately, both sides have it wrong. Recycling is neither "good" nor "bad"; solid waste is neither trash nor treasure. Like all other activities, recycling makes economic and environmental sense in some cases and not in others. The challenge is to figure out how to tell which cases are which—not to describe recycling in a simplistic, catch-all sound bite.

During the past decade, responding to public concern about resource conservation, most states adopted either "supply side" or "demand side" policies to encourage recycling. (Consumers who recycle are the suppliers of recyclables; firms who use recycled materials are the demanders.) Forty-one states went the supply-side road, adopting waste-diversion or recycling laws. Attempts to enact demand-side policies were less successful. Only two states—Oregon and California—passed recycled-content mandates for products other than newsprint. Florida passed (and then allowed to sunset) an "advance disposal fee," which is a tax on non-recyclable packaging. Nevertheless, since the prices of recyclables vary greatly, when scrap values for recyclables fell in the mid-1990s—many recyclable values are at near-historic lows today—the press for policies intended to stimulate demand resurfaced. Legislators and recycling advocates in Massachusetts, New York, Georgia, and elsewhere began pushing for new recycling legislation. The Environmental Protection Agency (EPA) is also considering raising its national recycling goal from 25% to 35%.

The premise, stated or unstated, behind these initiatives is that recycling reduces waste. This may seem obvious—at least if we use the intituitive definition that waste means "throwing things away instead of reusing them." But this is proof by definition. If we adopt a more sensible definition of waste—"doing with more resources what can be done with fewer resources"—the picture becomes more complicated. Whether we use virgin or recycled materials, manufacturing a product always uses resources of some sort, and it is hard to say ahead of time which method uses more resources. Is virgin manufacturing more wasteful than recycled manufacturing? It is really an empirical question.

The correct answer is: "It depends." Certain forms of recycling have existed for ages, because they save resources. Recycling aluminum conserves 95% of the energy required to make primary aluminum, and doesn't change the physical properties of the metal. Using one ton of recycled aluminum avoids the use of 4 tons of bauxite and 700 kg of petroleum coke and pitch, and avoids the emission of 35 kg of aluminum fluoride. But not all products are like aluminum. Some bath tissues and paper towels can be made out of 100% recycled content, but they aren't as soft or absorbent, and have lower wet strength and use more fiber than virgin-material products. As a result, people who use the recycled product use more towels at a time. Ironically, using recycled materials may not even reduce total solid-waste generation in this case.

"Recycling is neither 'good' nor 'bad'; solid waste is neither trash nor treasure."

Or consider coffee "brick packs." Made out of an aluminum foil/plastic laminate, these aren't easily recyclable and can't be made from recycled content. But they are also lightweight and produce only 15% as much waste as traditional metal cans. Steel cans are easily recyclable, and are in fact recycled at a rate higher than 50%. But unless steel can recycling rates rise to 85%—which is highly unrealistic—the non-recyclable brick packs actually take up less landfill space. The same paradoxes can be illustrated using other materials—recycled vs. virgin plastic grocery bags; glass juice bottles vs. aseptically packaged juice boxes. The fact is that reality is tricky. Resource use is complex—more complex than most recycling advocates realize. Intuition suggests that recycling always reduces landfill usage and other resource use, but this intuition is sometimes wrong.

Does Recycling Make Sense?

Everything seems more convincing when attached to precise numbers. To precisely discuss whether recycling "makes sense" in the aggregate, we would have to gather a mind-numbing array of statistics. For any firm, the costs of producing from virgin or recycled materials depend on a number of factors. How scarce are resources, now and in the future? What sorts of machinery is the firm using currently, and would switching from virgin to recycled production require buying new equipment? What sort of market is there for recycled products? How easy will it be to get recycled materials of the proper quality?

For any consumer, the benefits of using virgin or recycled materials also depend on a number of factors: How much will a product cost if it's made from virgin or recycled materials? Are there quality differences, and if so, how important are they?

> ***"Ironically, using recycled materials may not even reduce total solid-waste generation."***

The answers to these questions are different for each firm and each consumer. Changing local conditions will change the relative costs of using virgin or recycled materials. Moreover, much of this information isn't written down or even observable. It exists in the heads of individual business people and consumers, who aren't required to reveal their costs or preferences, and who may have trouble making their tacit knowledge and intuition explicit even if they wanted to. All of which means that any effort to discuss whether recycling "makes sense" in the aggregate should be taken with a large shaker of salt. Still, we can engage in the following intellectual exercise. Suppose every company has identical costs, equal to national averages. Suppose all packaging material can be recycled back into the same sort of packaging material (of equal quality), which consumers will still buy. Finally, suppose each industry—glass packaging, paper packaging, steel packaging, and different varieties of plastic packaging—stays the same size over time, and that there is no substitution between materials. These are highly stylized assumptions, but without them, we can't even begin to talk about aggregate costs and benefits of recycling. (On average, our assumptions stack the deck in favor of recycling.)

Now, we can calculate the average cost of making all-virgin packaging, and compare it with the average cost of making packaging with a given level of recycled materials (say, 30%). The difference between those two costs is the average cost (or average benefit) of using that level of recycled content. To be truly comparable, of course, these costs should incorporate all elements of the product's life-cycle. For virgin materials, we consider the costs of extraction, production, and landfilling. For recycled materials, we consider the costs of recyclables collection, production using recycled content, and reintroduction of the materials into the recycling stream. All costs include energy and transportation costs; the costs of recycled materials include estimates of how much waste can be reasonably collected and recycled, how much is lost during the production process, and how much will realistically end up being landfilled anyway. Finally, for an extra dose of realism, we estimate costs for a "best-case" scenario (where recycling is cheaper) and a "worst-case" scenario (where recycling is more expensive).

"As the amount of recycled content rises . . . the cost of using recycled materials rises, resulting in net societal losses when mandated."

Crunch the numbers, and we find that under best-case conditions, using some recycled content produces net benefits to society for almost all materials—paper, glass, metal, and plastic. For example, under best-case conditions, making glass packaging with 30% recycled content can yield (in the aggregate, across all production) benefits of $4 per ton over making glass packaging out of virgin material, and making paper with 30% recycled content can yield benefits of $50 per ton. But as the amount of recycled content rises and conditions become less favorable, the cost of using recycled materials rises, resulting in net societal losses when mandated. For instance, under worst-case conditions, requiring 30% recycled content in all glass packaging can cost—again, in the aggregate, across all production—$119 per ton more than using virgin material, and requiring that paper contain 30% recycled content costs $80 more per ton.

Remember that the numbers used in this model are approximate national averages. Actual benefits and costs for different manufacturers—and therefore, the best amount of recycled content to use—vary widely. While these numbers give us a rough sense of aggregate costs and benefits, we shouldn't believe in them too deeply; they may not actually be true of any individual manufacturer. Because all manufacturers are different, the best mandate is no mandate. Forcing specific, arbitrary levels of recycling will be counterproductive.

The Problem with Government Recycling Policies

While forcing specific, arbitrary levels of recycling may be counterproductive, it is precisely what many governments around the world have done. Recycled-content mandates—"Thou shalt use x percent recycled content"—are a common and obvious way of increasing the use of recycled materials, though like all one-size-fits-all resource-use plans, they make about as much sense as "Thou shalt use steel." Then there are virgin material taxes, designed to discourage the use of virgin materials by making them more expensive relative to recycled materials. Again, such taxes may make sense if recycling always saved resources, but in the complicated world we actually live in, they are unlikely to produce environmental benefits if applied across the board.

Other recycling-friendly policies are more complicated, but share many of the same failings.

• *Disposal and Recycling Fees.* Florida has tried a system of "advance disposal fees" (ADFs). Essentially, ADFs are a tax, added to the cost of a product, that incorporates the cost of disposing of that product when it is thrown away. Because poor families spend a higher proportion of their income on consumer goods, ADFs are a regressive tax; some of the proposed higher-end ADFs (around 10 cents per package) could be equivalent to a sales tax increase of as much as 7.5%.

The Florida government determined that ADFs weren't an effective way to promote efficient resource use. ADFs didn't increase recycling much; recycling did increase in Florida, but it also increased nationwide, and most of the increase predates the adoption of ADFs. On the other hand, while ADFs cost Florida consumers $45-$50 million, only $6 million of that was spent on recycling-market development. In the end, though, it was the complexity of the real world that made ADFs unworkable.

In practice, it is impossible for governments to come up with accurate and efficient fees for individual products. Disposal and recycling costs are location-specific; there is no uniformly applicable national or statewide cost. Recycling costs vary both by material and by product type; setting fees by material obscures differences among products, while setting fees by product can involve thousands of different items. Collection and disposal costs vary both by weight and by volume, which further complicates "efficient" fee calculations. And finally, disposal costs are dynamic. They change over time, sometimes rapidly—definitely far faster than government agencies can respond.

"One-size-fits-all resource-use plans . . . make about as much sense as 'Thou shalt use steel.'"

• *Packaging Take-Back Programs.* German waste-reduction policies have been even more innovative than most U.S. recycling efforts. In 1991, Germany adopted a system of "manufacturers' responsibility," commonly called the "Green Dot" system, in which manufacturers are required to "take back," or otherwise implement waste-recovery and material-recovery systems, for their packaging. Manufacturers who participate in the system, for a fee, put a green dot on their packaging. This packaging can be thrown out in special green-dot bins, and is collected for recycling by an industry-funded collection-and-recycling system.

It isn't clear that the German system has been greatly effective. Reductions in packaging use were about the same in Europe as in the United States, which had no such system—from over 2,500 pounds per gross production unit in 1989 to about 2,100 in 1994. On the other hand, the program has been expensive. Maintaining two separate waste-collection systems—regular trash cans, and green-dot bins—is costly. Improper disposal is widespread; as much as 40% of packaging in the special bins is regular trash. Also, the costs of recycling some of the materials have been high—two to three times U.S. recycling costs.

Manufacturers' responsibility (or "product stewardship") programs have arisen in the United States through market forces for certain products, such as solvents, pesticides, herbicides, and other products with high potential toxicity if improperly handled and disposed of. In these instances, manufacturers have an incentive to take back these discarded products after use, or simply to lease rather than sell them to consumers, to avoid any potential liabilities that might occur when consumers improperly handle and dispose of these products. Film-processing companies, computer manufacturers, and other makers of specialty products that have either high recycling value or pose a significant hazard if improperly disposed of have also created their own private take-back programs. But most products don't fit this mold. Where transactions are high-volume and low-value, where products are highly heterogeneous and widely distributed geographically, and where the waste-handling or product-management infrastructure is already safe and efficient, packaging take-back policies are hard to implement and enforce.

"'Recycling' isn't the opposite of 'waste.'"

Reality may be tricky, but there are some simple lessons to be learned from the recycling debate. The first is that "recycling" isn't the opposite of "waste." Sometimes recycling saves resources; sometimes it doesn't. Encouraging recycling across the board may reduce or increase energy use, water use, air and water pollution, and landfill usage—depending on the manufacturer and product. Given such variations, mandates for particular recycling rates or recycled-content levels are unlikely to create environmental benefits.

The Strength of the Price System

The second lesson is that we already have a way of figuring out when particular uses of resources make sense and when they don't, and this method doesn't involve mandates or government micromanagement. It's called market pricing.

If all resources—oil, wood, steel, landfill space, air and water quality, labor—were universally abundant, everyone would be able to consume whatever resources they wanted without limit. Prices emerge as a result of scarcity. When a resource becomes scarcer, its price increases; consumers, responding to this market signal, cut down on their use of the resource. When a resource becomes more abundant, its price drops, signaling to consumers that they can use more of it.

The price of a product, in a competitive economy, is, all in all, a pretty good indication of its resource-intensiveness. We can, therefore, roughly estimate that when recycling is more expensive than using virgin materials, this is probably because it uses more resources. The strength of the price system lies in its decentralization. Everyone in the economy knows the price of the products they buy, and generally tries to economize where possible, without any directives from a higher authority.

Of course, the price system isn't perfect. Many consumers don't have to pay for trash collection based on the weight or volume of trash they discard or, if they do, they often pay artificially low prices that don't reflect the actual cost of disposing of trash. (On the bright side, variable-rate pricing, also called "pay as you throw" garbage collection, is enjoying increased popularity in many cities.) The extraction of virgin materials, such as lumber or oil, is subsidized by the government, though the overall effects of these subsidies seem to be rather small. Such distortions to the price system, which make some resources appear more or less expensive than they actually are, should be fixed. But the problems of resource use will not be fixed by introducing new distortions: Advance disposal fees, for example, will skew, not improve, market signals about efficient resource use. Manufacturer take-back systems should be allowed to emerge voluntarily in the marketplace where they make sense.

"Is recycling good or bad?" then, is the wrong question to ask. Recycling can be beneficial, but sometimes isn't the best way to save resources. The real question is whether government mandates are a good way to foster resource conservation. The spotty experience of solid-waste legislation suggests that the marketplace does a better job of fostering recycling where it makes sense.

Recycling Does Not Reduce Waste

by Peter Werbe

About the author: *Peter Werbe writes articles on various cultural and political issues from a liberal point of view for his website, the Peter Werbe Article Database, found at http://goodfelloweb.com.*

The title of this article ["How I Stopped Recycling and Learned to Love It"] is somewhat misleading since I continue to recycle a portion of the waste produced daily by my household. What has changed is my previous diligence in making certain every scrap of what is recyclable winds up in my yellow and green curbside container.

Abandoning a "Good Faith" Solution

Now I use my recycle bin solely because my trash has to be placed somewhere for disposal. However, if I had to make any concerted effort at all, such as sorting or transporting my trash to some facility, I'm sure I wouldn't bother.

I realize even the headline is a provocation to some people who see recycling as an important component in the campaign for a clean environment. However, the case can be made that not only is this an inadequate perspective, it leads eventually to the *opposite* of its intent.

Recycling is a classic case of co-optation by the reigning powers of genuine sentiment for reform. The idea of reprocessing waste items was put forth as a good faith solution by those in the ecology movement who saw the damage being done to the environment by the detritus of production and consumption.

Recycling also gained impetus in the 1980s as an alternative to the rash of huge incinerators being built such as one in Detroit. This monstrous facility, the world's largest at the time of construction, sits three miles from the downtown area, less than a mile from a middle school in the midst of a poor African-American district. This insane techno-fix (doesn't everyone know burning *anything* produces toxins?) has as its basis the idea that we can continue current waste levels without having to pay the consequences.

Any sort of conservation or recycling is officially discouraged since these babies need all the fuel they can get, often to meet contract requirements with local utilities to produce electricity. Unfortunately for the environment and the people living in the immediate area (almost always poor and/or minority), these incinerators emit a deadly stream of dioxins, furans, and heavy metals into the air which assault our immune system.

Even with all the evidence about toxicity levels emanating from incinerators, their fires remain stoked, and they continue to produce toxic ash (as much as 30 percent of what is burned needs to be buried in special landfills to contain their now-concentrated poisonous content).

Economically, incinerators are flatlining all over the country due to their inability to produce the electricity for which they contracted to utilities. At one Detroit area burner, the operating authority has set up a special marketing division to seek trash from surrounding municipalities, even Canada if necessary, to meet its fuel needs.

Recycling: Better than Nothing?

In contrast, recycling seems like a reasonable alternative, particularly since it doesn't confront either our personal consumption level or society's aggregate mess. The only demand is that people place recyclables in a separate bin, something with which most good citizens were willing to comply even when not required by local ordinance. In municipalities where curbside recycling isn't provided as a city service, many people willingly make trips to recycling centers with their sorted trash feeling "they're at least doing *something*."

However, the "something" is illusory. Even with all the tonnage being recycled, landfills remain the major destination for the majority of household garbage and when space runs out like it has at New York's Fresh Kills facility, the city contracts to have it shipped to sites in Virginia.

A quick visual check in your neighborhood should illustrate that recycling isn't significantly reducing the trash that will either be landfilled or incinerated. Estimate the volume in the non-recycled section of your trash or on your block compared to the relatively tiny amount in recycling containers. My box is filled maybe every two weeks, much of it with newspapers, but every week I set out one or two 30-gallon garbage cans. And that's with at least some consciousness on my part about waste, excessive consumption and the composting of all my vegetable matter.

> ***"Recycling is a classic case of co-optation by the reigning powers of genuine sentiment for reform."***

Some people argue that if recycling is not effective, it at least functions as a gesture and is an important element towards understanding individual responsibility for our mess. The notion, however, that recycling is even a little better than nothing produces only more illusions, not environmental sanity. Mad levels of production and consumption are at the core of market economies, and unless that process is confronted, little will change.

To some extent this viewpoint about individual disposal of household garbage should only be a footnote when talking about waste. Americans generate 8.5 billion tons of waste yearly, but the vast majority—98 percent—is from industrial and mining operations. The remaining two percent—172 million—is from municipal sources. According to the Summer 1990 *Earth Island Journal,* the latter totals out to an average of 1360 pounds per person yearly for households, but a whopping 31 tons (!) for each of us from the major sources. The emphasis on household recycling functions as a diversion from examining the big sources of waste.

The Problem with Plastic

A close look at the myths about recycling shows they are being perpetrated less by those committed to ecology and more by those doing the most damage to the planet. Even those active in administering recycling programs have come to recognize, for instance, that plastics consumption (an increasing percentage of the waste stream) is actually *encouraged* by recycling. For that reason, the Berkeley Ecology Center (BEC) announced in February 1996 that it would no longer accept plastics in the recycling program they administer for that California city.

"Even with all the tonnage being recycled, landfills remain the major destination for the majority of household garbage."

Though they don't use it in production, the American Plastics Council, an industry group for virgin resin manufacturers (first-time-use plastics) has been a relentless promoter of plastics recycling. They've recently spent $18 million on public relations as part of a propaganda campaign to change the long-standing perception of their product as harmful to the environment.

From its inception plastic has been a synonym for the false and insubstantial. The late Frank Zappa sang about "plastic people" and the obscenely whispered advice to [Dustin Hoffman's character in the film] "The Graduate," similarly was, "Plaaastics." Unfortunately, the businessman in the 1967 film ultimately was correct; the future did lie in that multi-use substance made from the oil for which the U.S. is willing to kill hundreds of thousands of Iraqis. The substitution of plastics for glass, wood and paper products in daily use has been so substantial that hardly anyone remembers the difference from only a few decades ago. At any public event, for instance, such as a baseball game, massive amounts of plastic cups, plates and cutlery are used, in some cases for only the seconds it takes to spill down ten ounces of beer before being consigned to a trash barrel.

The cups arrive at the local landfill (they can't be recycled), there to remain intact for hundreds of years, although their slow disintegration begins to release toxins. They began their ignominious journey in an oil field thousands of miles away and were toxic every moment of their existence—from drilling to oceanic transportation, to off-loading at American harbors to manufacture and finally to disposal. Plants that pump out benzene and vinyl chlorines, building blocks for a wide spectrum of plastics, produce 14 percent of U.S. toxic air emissions. Sixteen percent of all industrial accidents—explosions, toxic cloud releases, chemical spills and fires—involve plastic production. Recycling doesn't touch this, but the spills and accidents aren't what are featured in industry ads.

"Recycled plastic is a small percentage of what is manufactured and the amount is actually **decreasing."**

Recycled plastic is a small percentage of what is manufactured and the amount is actually *decreasing* even as recycling increases. In 1993, for instance, 15 billion pounds of plastic were produced from what the industry calls virgin feed stock, but only one billion pounds of that was recycled.

And, the "at least we're doing something" argument doesn't work well here either. The industrial process which reclaims plastic is highly toxic and much of what is collected is shipped overseas, and processed under uncontrolled conditions in notorious polluting countries like China and Thailand. In addition, most of the products which are manufactured from what is recycled, such as park benches, traffic strips, and polyester jackets, can't be recycled a second time. So, what you set out at your curb is only one generation away from a landfill.

Michael Garfield, director of the Ann Arbor (Michigan) Ecology Center, notes that although all plastic containers bear the chasing arrows symbol with a number in the middle, suggesting that all such products are recyclable, it is only 1s and 2s that can be. He says, "Recycling these are only slightly better than letting them go into a landfill, given the amount of resources expended."

He's being generous if you compute the energy needed to ship your leftover designer water bottle to China along with millions of others to be reprocessed, manufactured into a new item, then shipped back to the U.S., transported to a mall, purchased, used, discarded, and finally landfilled.

It's interesting to note how the last imperative in the ecological triad of reduce, reuse, recycle, has emerged as the one given prominence. The consequences of demanding an emphasis on the first—reduction of consumption and production—puts one on the path of confrontation with a system which is seeking to function at maximum capacity, not minimum.

A good ecologist may now need to do more than just put tin cans in a curbside recycling bin. For instance, a campaign against plastic demands opposition not only to oil as a world commodity, but also to what countries depending on it are willing to do to control it.

Organizations to Contact

The editors have compiled the following list of organizations concerned with the issues debated in this book. The descriptions are derived from materials provided by the organizations. All have publications or information available for interested readers. The list was compiled on the date of publication of the present volume; the information provided here may change. Be aware that many organizations take several weeks or longer to respond to inquiries, so allow as much time as possible.

American Chemistry Council
1300 Wilson Blvd., Arlington, VA 22209
(703) 741-5000 • fax: (703) 741-6000
e-mail: webmaster@americanchemistry.com • website: www.cmahq.com

The council represents the U.S. chemical industry on public policy issues, coordinates research and testing programs, and administers the industry's environmental, health, and safety performance improvement initiative, known as Responsible Care. It asserts that the chemical industry is actively engaged in improving the environmental safety of its products and publishes the *Responsible Care News* and the *LRI Update* newsletter, both on a monthly basis.

American Council on Science and Health (ACSH)
1995 Broadway, 2nd Floor, New York, NY 10023-5860
(212) 362-7044 • fax: (212) 362-4919
e-mail: acsh@acsh.org • website: www.acsh.org

ACSH is a consumer education consortium concerned with, among other topics, issues related to the environment and health. The council publishes the quarterly *Priorities* magazine and position papers such as "Global Climate Change and Human Health" and "Public Health Concerns About Environmental Polychlorinated Biphenyls."

Canadian Centre for Pollution Prevention (C2P2)
100 Charlotte St., Sarnia, ON, N7T 4R2 Canada
(800) 667-9790 • fax: (519) 337-3486
e-mail: info@c2p2online.com • website: http://c2p2online.com

The Canadian Centre for Pollution Prevention is Canada's foremost pollution prevention resource. It offers easy access to national and international information on pollution and prevention through a search service, hard copy distribution, an extensive website, online forums, publications, and customized training. Among their publications are the *Practical Pollution Training Guide* and *at the source*, C2P2's newsletter produced three times a year.

Cato Institute
1000 Massachusetts Ave. NW, Washington, DC 20001-5403
(202) 842-0200 • fax: (202) 842-3490
e-mail: cato@cato.org • website: www.cato.org

The Cato Institute is a libertarian public policy research foundation dedicated to limiting the role of government and protecting individual liberties. It disapproves of EPA regulations, considering them to be too stringent. The institute publishes the quarterly magazine *Regulation* and the bimonthly *Cato Policy Report*, as well as over one hundred papers dealing with the environment including "Why States, Not EPA Should Set Pollution Standards" and "The EPA's Clean Air-ogance."

Earth Systems
508 Dale Ave., Charlottesville, VA 22903
(804) 293-2022
e-mail: www@earthsystems.org • website: www.earthsystems.org

Earth Systems is a nonprofit organization that develops, compiles, categorizes, and delivers environmental education and information resources to the world at large. It offers a virtual library with an index of over nine hundred online environmental resources and *EcoTalk*, a mailing list devoted to helping nonprofit environmental organizations find solutions to problems.

Environmental Industry Associations (EIA)
4301 Connecticut Ave. NW, Suite 300, Washington, DC 20008
(202) 244-4700 • fax: (202) 966-4818
e-mail: eia@envasns.org • website: www.envasns.org

Affiliated with the National Solid Wastes Management Association and the Waste Equipment Technology Association, EIA represents about two thousand businesses that manage solid, hazardous, and medical wastes; manufacture, distribute, and service waste equipment; and provide environmental management and consulting services. It publishes the newsletter *Infectious Wastes News* and offers several pamphlets and profiles on various waste management issues.

Environmental Protection Agency (EPA)
401 M St. SW, Washington, DC 20460-0001
(202) 382-2090
website: www.epa.gov

The EPA is the federal agency in charge of protecting the environment and controlling pollution. The agency works toward these goals by assisting businesses and local environmental agencies, enacting and enforcing regulations, identifying and fining polluters, and cleaning up polluted sites. It publishes the monthly *EPA Activities Update* and numerous periodic reports.

Environmental Research Foundation (ERF)
PO Box 5036, Annapolis, MD 21403-7036
(410) 263-1584 • fax: (410) 263-8944
e-mail: erf@rachel.org • website: www.rachel.org

The foundation works with various groups and individuals, such as grassroots activists and public-interest scientists, to motivate communities to act against the dangers of all types of pollution. It specializes in information about hazardous waste and waste disposal and seeks to educate the public on their adverse health effects. The foundation's publications include the biweekly newsletter *Rachel's Environment and Health News* and numerous reports such as *Dangerous Substances in Waste* and *American People's Dioxin Report.*

Foundation for Clean Air Progress (FCAP)
1801 K St. NW, Suite 1000L, Washington, DC 20036
(800) 272-1604
e-mail: info@cleanairprogress.org • website: www.cleanairprogress.org

FCAP is a nonprofit organization that believes the public remains unaware of the substantial progress that has been made in reducing air pollution. It represents various sectors of business and industry in providing information to the public about improving air quality trends. In support of its call for less government regulation, FCAP publishes numerous reports and studies demonstrating that air pollution is on the decline.

Friends of the Earth
1025 Vermont Ave. NW, Washington, DC 20005
(202) 783-7400 • fax: (202) 783-0444
e-mail: foe@foe.org • website: www.foe.org

Friends of the Earth is a national advocacy organization dedicated to protecting the planet from environmental degradation; preserving biological, cultural, and ethnic diversity; and empowering citizens to have an influential voice in decisions affecting the quality of their environment. It publishes the quarterly *Friends of the Earth Newsmagazine* and *Atmosphere*, a report focusing on actions taken to preserve the ozone layer published three times a year.

Heritage Foundation
214 Massachusetts Ave. NE, Washington, DC 20002-4999
(800) 546-2843 • (202) 546-4400 • fax: (202) 546-8328
e-mail: pubs@heritage.org • website: www.heritage.org

The Heritage Foundation is a conservative think tank that supports free enterprise and limited government. Its researchers criticize EPA overregulation and believe that recycling is an ineffective method of dealing with waste. Its publications, such as the quarterly *Policy Review*, include studies on the uncertainty of global warming and the greenhouse effect.

INFORM
120 Wall St., New York, NY 10005-4001
(212) 361-2400 • fax: (212) 361-2412
e-mail: brown@informinc.org • website: www.informinc.org

INFORM is an independent research organization that examines the effects of business practices on the environment and on human health. The collective goal of its members is to identify ways of doing business that ensure environmentally sustainable economic growth. It publishes the quarterly newsletter *INFORM Reports* and fact sheets and reports on how to avoid unsafe uses of toxic chemicals, protect land and water resources, conserve energy, and safeguard public health, many of which are available on the Internet.

National Recycling Coalition (NRC)
1727 King St., Suite 105, Alexandria, VA 22314
(703) 683-9025 • fax: (703) 683-9026
e-mail: info@nrc-recycle.org • website: www.nrc-recycle.org

The National Recycling Coalition is a nonprofit organization that promotes recycling as a way to reduce solid waste, protect the environment, and conserve resources. NRC provides information to the public on recycling issues and operates programs that encourage recycling markets and economic development. It publishes the quarterly newsletter the *Connection*, the monthly newsletter *Recycling Policy Reporter*, and reports in defense of recycling.

Natural Resources Defense Council (NRDC)
40 W 20th St., New York, NY 10011
(212) 727-2700 • fax: (212) 727-1773
e-mail: nrdcinfo@nrdc.org • website: www.nrdc.org

The Natural Resources Defense Council is a nonprofit organization that uses law, science, and more than four hundred thousand members nationwide to protect the planet's wildlife and wild places and to ensure a safe and healthy environment for all living things. NRDC publishes the quarterly magazine *OnEarth* in addition to hundreds of reports, including *Cool It: Eight Great Ways to Stop Global Warming* and *After Silent Spring: The Unsolved Problems of Pesticide Use in the United States.*

Pacific Northwest Pollution Prevention Resource Center (PPRC)
513 First Ave. W, Seattle, WA 98119
(206) 352-2050 • fax: (206) 352-2049
e-mail: office@pprc.org • website: www.pprc.org

PPRC is a nonprofit organization dedicated to preventing pollution in the Pacific Northwest. It organizes and publicizes conferences and seminars for individuals concerned about pollution prevention. PPRC publishes the bimonthly *Pollution Prevention Northwest* newsletter.

Political Economy Research Center (PERC)
502 S. 19th Ave., Suite 211, Bozeman, MT 59718
(406) 587-9591
email: perc@perc.org • website: www.perc.org

PERC is a nonprofit research and educational organization that seeks market-oriented solutions to environmental problems. Areas of research covered in the PERC Policy Series papers include endangered species, forestry, fisheries, mines, parks, public lands, property rights, hazardous waste, pollution, water, and wildlife. PERC conducts a variety of conferences, offers internships and fellowships, provides environmental education materials, and publishes the quarterly newsletter *PERC Reports.*

Reason Foundation
3415 S. Sepulveda Blvd., Suite 400, Los Angeles, CA 90034
(310) 391-2245 • fax: (310) 391-4395
e-mail: gpassantino@reason.org • website: www.reason.org

The foundation promotes individual freedoms and free-market principles. Its researchers do not believe that recycling is beneficial and contend that the dangers of ozone depletion and global warming are myths. It publishes the monthly magazine *Reason.*

Sierra Club
85 Second St., Second Floor, San Francisco, CA 94105-3441
(415) 977-5500 • fax: (415) 977-5799
e-mail: information@sierraclub.org • website: www.sierraclub.org

The Sierra Club is a grassroots organization with chapters in every state. Maintaining separate committees on air quality, global environment, and solid waste, among others, it promotes the protection and conservation of natural resources. It publishes the bimonthly magazine *Sierra* and the *Planet* newsletter, which appears several times a year, in addition to books and fact sheets.

Worldwatch Institute
1776 Massachusetts Ave. NW, Washington, DC 20036-1904
(202) 452-1999 • fax: (202) 296-7365
e-mail: worldwatch@worldwatch.org • website: www.worldwatch.org

Worldwatch is a nonprofit public policy research organization dedicated to informing policy makers and the public about emerging global problems and trends and the complex links between the world economy and its environmental support systems. It publishes the bimonthly *World Watch* magazine, the Environmental Alert series, and several policy papers.

Bibliography

Books

Rick Abraham — *The Dirty Truth: The Oil and Chemical Dependency of George W. Bush.* Houston, TX: Mainstream, 2000.

Harvard Ayres, Jenny Hager, and Charles E. Little — *An Appalachian Tragedy: Air Pollution and Tree Death in the Eastern Forests of North America.* San Francisco: Sierra Club, 1998.

Charles Bartsch and Elizabeth Collaton — *Brownfields: Cleaning and Reusing Contaminated Properties.* Westport, CT: Praeger, 1997.

Pamela S. Chasek, ed. — *The Global Environment in the Twenty-First Century: Prospects for International Cooperation.* Tokyo: United Nations University Press, 2000.

Nicholas P. Cheremisinoff — *Handbook of Pollution Prevention Practices.* New York: M. Dekker, 2001.

Terry Dinan and Christian Spoor — *An Evaluation of Cap-and-Trade Programs for Reducing U.S. Carbon Emissions.* Washington, DC: Congress of the United States, Congressional Budget Office, 2001.

Jack Doyle — *Taken for a Ride: Detroit's Big Three and the Politics of Pollution.* New York: Four Walls Eight Windows, 2000.

James A. Dunn Jr. — *Driving Forces: The Automobile, Its Enemies, and the Politics of Mobility.* Washington, DC: Brookings Institution Press, 1998.

Elizabeth Glass Geltman — *Recycling Land: Understanding the Legal Landscape of Brownfield Development.* Ann Arbor: University of Michigan Press, 2000.

Indur Goklany — *Clearing the Air: The Real Story of the War on Air Pollution.* Washington, DC: Cato Institute, 1999.

Hugh S. Gorman — *Redefining Efficiency: Pollution Concerns, Regulatory Mechanisms, and Technological Change in the U.S. Petroleum Industry.* Akron, OH: University of Akron Press, 2001.

Wyn Grant, Anthony Perl, and Peter Knoepfel, eds. — *The Politics of Improving Urban Air Quality.* Northampton, MA: Edward Elgar Publishing, 1999.

James D. Johnston — *Driving America: Your Car, Your Government, Your Choice.* Washington, DC: AEI Press, 1997.

Jane Holtz Kay — *Asphalt Nation: How the Automobile Took over America, and How We Can Take It Back.* New York: Crown Publishers, 1997.

Bjorn Lomborg — *The Skeptical Environmentalist: Measuring the Real State of the World.* Cambridge, England: Cambridge University Press, 2001.

Thomas H. Lynn — *The Hidden Dangers in Tap-Water and the Best Solutions.* Houston: Longevity Press, 1996.

James S. Lyon et al. — *Golden Dreams, Poisoned Streams: How Reckless Mining Pollutes America's Waters, and How We Can Stop It.* Washington, DC: Mineral Policy Center, 1997.

Gordon McGranahan et al. — *The Citizens at Risk: From Urban Sanitation to Sustainable Cities.* Sterling, VA: Earthscan, 2001.

Cassandra Chrones Moore — *Haunted Housing: How Toxic Scare Stories Are Spooking the Public out of House and Home.* Washington, DC: Cato Institute, 1997.

Klaus Nurnberger — *Prosperity, Poverty and Pollution: Managing the Approaching Crisis.* New York: Zed Books, 1999.

Gunter Pauli, J. Hugh Faulkner, and Fritjof Capra — *Upsizing: The Road to Zero Emissions, More Jobs, More Income and No Pollution.* Sheffield, England: Greenleaf, 2000.

Paul R. Portney and Robert N. Stavins, eds. — *Public Policies for Environmental Protection.* Washington, DC: Resources for the Future, 2000.

Joe Schilling, Christine Gaspar, and Nadejda Mishkovsky — *Beyond Fences: Brownfields and the Challenges of Land Use Controls.* Washington, DC: International City/County Management Association, 2000.

Robert A. Simons — *Turning Brownfields into Greenbacks: Developing and Financing Environmentally Contaminated Urban Real Estate.* Washington, DC: Urban Land Institute, 1998.

Marvin S. Soroos — *The Endangered Atmosphere: Preserving a Global Commons.* Columbia: University of South Carolina Press, 1997.

Julie Stauffer — *The Water Crisis: Constructing Solutions to Freshwater Pollution.* Tonawanda, NY: Black Rose Books, 1999.

Susan Strasser — *Waste and Want: A Social History of Trash.* New York: Metropolitan Books, 1999.

Jacqueline Vaughn Switzer — *Environmental Politics: Domestic and Global Dimensions.* Boston: Bedford/St. Martin's, 2001.

James M. Symons — *Plain Talk About Drinking Water: Questions and Answers About the Water You Drink.* Denver: American Water Works Association, 1997.

Martyn Turner and Brian O'Connell — *The Whole World's Watching: Decarbonizing the Economy and Saving the World.* Chichester, England: John Wiley & Sons, 2001.

John Wargo — *Our Children's Toxic Legacy: How Science and Law Fail to Protect Us from Pesticides.* New Haven, CT: Yale University Press, 1998.

Adam S. Weinberg, David N. Pellow, and Allan Schnaiberg — *Urban Recycling and the Search for Sustainable Community Development.* Princeton, NJ: Princeton University Press, 2000.

David Wheeler — *Racing to the Bottom? Foreign Investment and Air Pollution in Developing Countries.* Washington, DC: World Bank, Development Research Group, Infrastructure and Environment, 2001.

B.C. Wolverton and John D. Wolverton — *Growing Clean Water: Nature's Solution to Water Pollution.* Picayune, MS: Wolverton Environmental Services, 2001.

World Bank, Development Research Group — *Greening Industry: New Roles for Communities, Markets, and Governments.* Washington, DC: World Bank, 2000.

Periodicals

Claire Barliant and Mike Burger — "Smokestack Lightning: Online Trade in Greenhouse Gases Strikes Environment," *Village Voice*, December 12, 2000.

Peter Barnes — "The Pollution Dividend," *American Prospect*, May 1999.

Virginia Warner Brodine — "Corporate System Is Main Threat to the Environment," *People's Weekly World*, September 13, 1997.

H. Sterling Burnett and Merrill Matthews Jr. — "Mr. Gore, Carbon Dioxide Is Not a Pollutant," *Human Events*, March 6, 1998.

Christian Science Monitor — "Cleaning Up Superfund," March 28, 1997.

Marla Cone — "Vehicles Blamed for a Greater Share of Smog," *Los Angeles Times*, October 30, 1999.

John H. Cushman Jr. — "Pollution Control Plan Views Factory Farms as Factories," *New York Times*, March 6, 1998.

James Dao — "Study Sees Acid Rain Threat in Adirondacks and Beyond," *New York Times*, April 5, 1999.

Claudia H. Deutsch — "Together at Last: Cutting Pollution and Making Money," *New York Times*, September 9, 2001.

Economist — "Our Durable Planet: Why Poor Countries Are More Polluted," *San Diego Union-Tribune*, November 28, 1999.

John L. Fialka — "EPA Study to Call for Tougher Controls on Emissions of Automobiles by 2004," *Wall Street Journal*, April 23, 1998.

Carl Frankel — "Clean Water: What's It Worth?" *Yes!*, Summer 1998.

Gary M. Galles — "Did the EPA Overstep Its Powers?" *San Francisco Chronicle*, May 25, 2000.

Wendy L. Gramm and Susan E. Dudley — "The Human Costs of EPA Standards," *Wall Street Journal*, June 9, 1997.

Kenneth Green — "Weighing Fairness and Pollution Regs for Sport Utility Vehicles," *San Diego Union-Tribune*, November 5, 1998.

William Greider — "Carbongate," *Amicus Journal*, Summer 2001.

Jane Holtz Kay — "Car Sick Country," *Sierra*, July 1999.

Angus King and Ned Sullivan — "Too Many Bad Air Days," *New York Times*, August 25, 1999.

Margaret Kriz — "EPA Lights a Coal Fire," *National Journal*, January 29, 2000.

Susan Lee — "How Much Is the Right to Pollute Worth?" *Wall Street Journal*, August 1, 2001.

Matthew Leising — "Trading Gas," *Village Voice*, August 29, 2000.

Ben Lieberman — "Air Pollution—The Inside Story," *Regulation*, Spring 1998.

Vernon Masayesva — "We Can Have Electricity, Jobs and Clean Air," *High Country News*, March 30, 1998.

Lisa Mastny — "Ozone Hole Is Largest Ever," *World Watch*, January/February, 1999.

Kim Murphy — "Arctic Oil Pipeline Brings Out Activists, Supporters," *Los Angeles Times*, April 29, 2000.

New York Times — "New Threats to Cleaner Air," May 26, 2001.

New York Times — "Pressuring the Power Plants," March 28, 2000.

Gary Polakovic — "Earth Losing Air-Cleansing Ability, Study Says," *Los Angeles Times*, May 4, 2001.

Gary Polakovic — "Smog Feared in Power Buildup," *Los Angeles Times*, July 16, 2001.

Phil Radford — "Bush's Hot Air," *Multinational Monitor*, May 2001.

April Reese — "Bad Air Days," *E*, November 1999.

Andrew C. Revkin — "Record Ozone Hole Refuels Debate on Climate," *New York Times*, October 10, 2000.

Bill Richards — "Exxon Is Battling a Ban on an Infamous Tanker," *Wall Street Journal*, July 29, 1998.

Fred Richardson and Andrew Wheat — "Dirty Old Grandfathered Plants: The Clean Air Act's Lung-Charring Loophole," *Multinational Monitor*, June 1998.

Michael A. Rivlin — "The Superstress of Superfund," *Amicus Journal*, Winter 1999.

Lori Saldana — "Tackling the Border Sewage Problem," *San Diego Union-Tribune*, October 14, 1998.

Kara Sissell — "Supreme Court Upholds EPA in Air Standards Dispute," *Chemical Week*, March 7, 2001.

Thomas Sowell — "Arsenic Scare Is Pure Politics, Not Science," *Human Events*, April 30, 2001.

William K. Stevens — "Enormous Haze Found over Indian Ocean," *New York Times*, June 10, 1999.

William K. Stevens — "Water: Pushing the Limits of an Irreplaceable Resource," *New York Times*, December 8, 1998.

Steve Voynick — "Living with Ozone," *World and I*, July 1999.

Peter Waldman — "All Agree Arsenic Kills; The Question Is How Much It Takes to Do So," *Wall Street Journal*, April 19, 2001.

Murray Weidenbaum — "Pollution-Reporting Law Doesn't Apply to Government Polluters," *Insight on the News*, October 26, 1998.

Index

Abzug, Bella, 153
acid rain
  coal use and, 33
  nitrogen oxides and, 48
  oil extraction in Niger Delta and, 88
Ackerman, Frank, 194
advance disposal fees (ADFs), 199
Agip, 87
agricultural land use, 132
agricultural nonpoint source pollution, 133–34, 135–36
air pollution
  coal use and, 64
  congressional action on, 18
  controlling
    indoor, 22–23, 24–25
    outdoor, 20–22
  corporations reducing emissions for, 101–102
  determining acceptable risk of, 159
  EPA standards for, 53, 137–39
  faulty risk assessment for, 159–62
  fuel efficiency and, 51–52
  improvement in, 54–58
  regulating. *See* Clean Air Act Amendments
  sources of, 23–24
    automobiles as, 45, 46, 47–50
    indoor, 22
    outdoor, 18–20
  toxic pollutants in, 58–59
  see also pollution
Air Pollution Control Act (1955), 18
Alaska, 45, 80, 106
Albemarle, 40–41
Alcoa, 99
aluminum cans, 93–94
American Council for an Energy-Efficient Economy, 44
American Petroleum Institute, 101
American Plastics Council, 105, 204
Ammons, Bert, 36
Andrews, Mary, 193
animal factories. *See* factory farms
Antarctic, 42
antisapstain chemicals, 125–26
Archer Daniels Midland (ADM), 79
Arctic, 42
asbestos, 24
Association of Metropolitan Sewerage Agencies (AMSA), 154
automobiles
  carbon monoxide emissions from, 57
  emissions from
    imperceptible, 75
    lead, 58
    projections on, 73
    reducing, 21
    standards for, 46, 72–73
  environmental benefits of, 70–71
  environment harmed by, 44–45
  fuel economy standards for, 46–47
  fuel efficiency and, 51–52
  global warming and, 45, 50–51, 73–75
  health benefits of, 71–72
  pollutants from, 45–46, 47–50
  recyclable materials from, 44
Aylward, David, 148

Baker, Linda, 36
Bast, Joseph L., 70
batteries (car), 44
Baucus, Max, 189–90
Bauen, Ausilio, 190
benzene, 49, 58
Berkeley Ecology Center, 204
beverage packaging, 94
Bhopal chemical leak, 77, 79
biologic contaminants, 24
Blatnik, John, 152
Boehlert, Sherwood, 145
Bowers Landfill, 122
Brough, Wayne T., 141
Browner, Carol, 60, 61, 109, 137, 142–43, 151
Buckley, James, 153, 154
Burnett, H. Sterling, 110
Bush, George W., 111

Business Council for Sustainable Development, 83
businesses. *See* companies

California Air Resources Board (CARB), 46
Canada
  factory farming in, 26–29
  wood preservation industry in, 124–28
Canadian Environmental Management Survey, 128
Canadian Environmental Protection Act (CEPA), 125
cancer
  overestimating risks of, 160–62
  skin, 38
Cape Cod National Sea Shore, 32
carbon dioxide
  coal use and, 34
  global warming and, 50–51
carbon monoxide, 20
  decreased, 57–58
  harm in, 50
  poisoning, 23
  sources of, 57
carbon sequestration, 69
cars. *See* automobiles
Carson, Rachel, 12
Carter, Jimmy, 115, 118
Cato Institute, 166, 174, 178, 179
Celan Coal Technology Program, 66
CFCs. *See* chlorofluorocarbons
chemical explosions, 79
chemical spills, 116
Chevron Corporation, 85, 87
Chinook Health Region, 29
Chisso Corporation, 79–80
chlorofluorocarbons, 82
  buying, 38–39
  illegal trade of, 36–37, 39–40
  phasing out use of, 37–38
Circleville, Ohio, 122
citric acid market, 79
Clairton, Pennsylvania, 161–62
Clapp, Philip E., 112, 113
Clean Air Act (1963), 18
Clean Air Act (1970), 18
  coal use and, 31
  goals of, 13
  human health comes before costs of, 139–40
  need to expand, 34
Clean Air Act Amendments (CAAA) (1990), 21
  costs, 157
  determining acceptable risk for, 159
  lack of demonstrable risks for, 157–58
Clean Coal Technology Program, 67, 68
Clean Power Act (2001), 34
Clean Smokestacks Act (2001), 34
Clean Water Act (1972)
  amendments to, 133–34
  formed in a state of panic, 152–53
  goals of, 13
  is ineffective, 151–52
  litigation and, 155
  unrealistic goals of, 153–54
Clinton, Bill, 110, 141
Clinton administration, 112–13
coal use
  acid rain and, 33
  combustion, 67
  demand for, 63–64
  global warming and, 34
  greenhouse gas emissions and, 65–66
  legislation for, 34–35
  pollution from, 31–32
  power plants for, 65
  prices, 63
  reducing environmental impact of, 66–69
  regulation of, 64–65
  smog formed from, 32
  soot formed from 32–33
  toxins formed by, 33–34
Cohen, Gary, 77
combustion by-products, 23
Community Right to Know and Emergency Planning Law, 77
companies
  chemical explosions and, 77, 79
  conservation by, 93–94
  as created by the citizenry, 78
  cutting emissions, 101–102
  deceiving the public, 82
  destroying rail system, 80
  energy management by, 102–103
  environmental record of, 93
  environmental spending by, 105–106
  on federal government regulation, 109–10
  job-related illness and injury rate in, 106–107
  lip service to environmental concerns by, 82–83
  Love Canal scandal, 96–97
  mercury poisoning and, 79–80
  methyl bromide and, 40–41
  Mexican border, preventing pollution, 98–100
  oil spills and leaks by, 80, 103–104
  polishing image of, 81–82

pollution reduction and, 94–96
price fixing by, 79
public relations campaigns, 83–84
public scrutiny and pressure on, 77
recycling and, 104–105, 184, 199–200
reuse, 186–87
threat to Niger Delta by, 85–92
utility, 65
see also factory farms
Competitive Enterprise Institute, 166
Comprehensive Environmental Response, Compensation, and Liability Act (1980), 118, 143
see also Superfund program
Comprehensive National Energy Strategy, 67–68
conservation, 104–105
Consumer Product Safety Commission, 22
Cooper, Mary H., 109, 164
Culbertson, Jim, 41
Cuyahoga River (Cleveland, Ohio), 12–13, 152

Dale, Ali, 81
Davies, Kert, 37, 43
Davies, Terry, 110–11
Dead Sea Bromine, 40–41
Delphi, 99
Denison, Richard A., 166
Department of Housing and Urban Development, 22
Destruction of Inner City Rail, 80
Deutsch, Claudia H., 77
diapers, reusable, 187
diesel engines, 47–48
diesel fuel, 45
see also oil (fuel)
Douglass, Christopher, 188
Duncan, John L., 93
Dupont, 82

Eagleton, Tom, 154
Earth Day, 13, 116
Easterbrook, Gregg, 70
E. coli, 27, 29
El-Airini, Omar, 38
electric trolley system, 80
electric vehicles, 45
Elf-Aquitaine, 87
Elizabeth, New Jersey, 116–17
emissions
estimates, 53–54
government regulation on, 111
greenhouse gas, 65–66, 102–103
see also global warming
imperceptible, 75
projections on, 73
reduction in, 94, 101
standards for, 46, 72–73
see also Clean Air Act Amendments
endangered species, 90
energy efficiency, 102–103
environment
automobile has a negative impact on, 44–45
automobile invention has benefited, 70–71
corporate lip service to, 82–83
Environmental Council of States, 21
Environmental Defense Fund (EDF), 60
environmental disasters
air pollution, 17–18
chemical explosions, 116–17
environmental movement, 12, 13–14
Environmental Protection Agency (EPA), 53, 98, 115
on agricultural land, 132
air pollution and, 18, 21, 22
air quality standards by, 53, 137–39
on emissions, 53–54, 101–102
factory farming and, 28, 30, 61
faulty risk assessment by, 159–62
formation of, 13
indoor air pollutants and, 22
on mercury, 34
on methyl bromide, 41
recycling and, 196
Safe Drinking Water Act and, 134
Superfund and, 120, 121, 142, 144–45, 146–48, 149
Environmental Quality Incentives Program (EQUIP), 136
Environment Canada, 125, 127, 128
Enviropig, 62
Essential Action, 85
Exxon Corporation, 80
Exxon Shipping, 80
Exxon Valdez spill, 45, 80

factory farms
disputes over threat of, 60–62
vs. family farms, 26
genetic engineering and, 62
human health and, 29
organizations opposing, 60
reforming, 30
size of, 26–27
threaten water quality, 27–29
waste contamination from, 16
Farm Bill (1996), 135–36

Federal Agricultural Improvement and Reform Act (1996), 135–36
Federal Energy Regulatory Commission, 64
Federal Water Pollution Control Act (1948), 133
Federal Water Pollution Control Act (1972). *See* Clean Water Act
Feedlot Alley, 26, 29
Firestone, 80
Floegel, Mark, 16
flue gas scrubbers, 66–67
formaldehyde, 23
fossil fuels. *See* coal use; emissions
Fraser River, 124–25, 128
Fraser River Action Plan (FRAP), 125
Friends of the Earth, 81–82
fuel economy standards, 46–47
fuel efficiency, 51–52
fuels. *See* oil (fuel)
furans, 127

Garfield, Michael, 205
gas flaring, 87–88
gasoline, 45
  see also oil (fuel)
gasoline storage tanks, underground, 103–104
gas stoves, 23
General Electric (GE), 13
General Motors (GM), 80
genetically modified foods, 81–82
genetic engineering, 62
Germany, 199–200
Gianessi, Leonard, 61
Gilpin, Richard, 81
Glickman, Dan, 61
Global Exchange, 85
global warming
  auto use and, 45, 50–51, 73–75
  coal use and, 34
  debate on government regulation of, 111–13
  ozone depletion and, 42–43
  risks of, 51
  see also greenhouse gas emissions
Goldstein, Bernard D., 159
Gore, Al, 12
government
  recycling policies by, 198–200
  see also legislation; regulations
Graham, John, 160
Graham, Mary, 12
Great Lakes Chemical, 40–41
greenhouse gas emissions, 65–66, 103
  carbon sequestrations and, 69
  see also global warming
Green Lights, Energy Star (EPA program), 103
ground-level ozone, 55

Halpern, Richard A., 61, 151
Harris, James, 83
Hasselback, Paul, 27
hazardous wastes. *See* toxic substances/pollution
health. *See* human health
Healthy People 2000, 25
Hershkowitz, Allen, 174
Highfill, Jannett, 194
Hooker Chemical Company, 96, 97
Hosford, William F., 93
human health
  air pollution's effect on, 18, 22
  asbestos and, 24
  automobile invention has benefited, 71–72
  biologic contaminants and, 24
  carbon monoxide and, 20
  combustion by-products and, 23
  factory farms and, 29
  global warming and, 51
  hydrocarbons and, 49
  Love Canal tragedy and, 115
  mercury and, 34
  nitrogen dioxide and, 20
  nitrogen oxides and, 48
  nonpoint source water pollution and, 132
  oil extraction in Niger Delta and, 89
  overestimating risks to, 160–62
  ozone depletion/pollution and, 19, 38, 138
  particulate matter and, 19, 47, 138
  radon and, 24
  smog and, 32
  soot and, 33
  sulfur dioxide and, 20
  tobacco smoke and, 23
  volatile organic compounds and, 23
  wood preservation facilities and, 124–25
Human Rights Watch, 87
Hyde, Jeffrey, 131
hydrochlorofluorocarbons (HCFCs), 37
hydrofluorocarbons (HFCs), 43, 82

illnesses. *See* human health
incinerators, 202–203
indoor air pollution, 22–25
industrial accidents/disasters, 77, 205
Industri-Plex, 121
Innovest Capital Risk Advisors, 95
Intergovernmental Panel on Climate

Change (IPCC), 34, 74–75
INTERPOL (International Criminal Police Organization), 79

Johnson Controls, 99
Jones, Robert, 152
Jorling, Thomas C., 116, 117
Juday, Dave, 60

Kemet, 99
Kennedy, Edward, 117
Klapatiuk, Les, 28
Kovacs, William L., 110
Krahn, Peter K., 124
Kripowicz, Robert S., 63
Kyoto Protocol (1997)
  costs, 112
  opposition to, 110, 111

Lake Snell Perry & Associates, 60
landfills
  costs, 167
  increase pollution, 168–69
  shortage of, 164
  space for, 189, 190
lawsuits, 65
Leach, Matthew, 190
lead, 18–19, 58
legislation
  air pollution control, 18
  coal use, 34–35
  factory farming, 28, 61–62
  hazardous waste, 116
  nonpoint source pollution, 135–36
  safe drinking water, 134–35
  see also Clean Air Act (1970); Clean Water Act; regulations; Superfund program
Lehr, Jay, 70
Leipzig Declaration, 74
Lewis, Jack, 14
Lichaa, Pierre, 98, 99, 100
Love Canal, 96–97, 115
Lovejoy, Stephen B., 131, 155
Lucas, Nigel J.D., 190
lysine market, 79

mangrove forests, 90
manure, 27–28
Mark and Spencer's (store), 82
Maryland, 61–62
McQuade, Jack, 36, 39
McQuiggie, Murray, 28
mercury
  coal and, 33
  poisoning, 79–80
  regulations, 65
methyl bromide, 37, 38, 40–42
Methyl Bromide Fairness Act, 40, 41
methylene chloride, 58
Meyer, Pamela, 17
Michel, Pascal, 29
Middle East, 46
Milwaukee, Wisconsin, 16
Minamata Bay, 79
Mineral Policy Center (MPC), 177
mining, 177–78
Mobil, 87
Mobre 4000, 164, 189–90
Mokhiber, Russell, 78
Monsanto, 81–82
Montreal Protocol of 1987, 36–38
Moody-Stuart, Mark, 84
motor oil, 105
MTBE (gasoline additive), 104
Multilateral Fund, 40
Muskie, Edmund, 152

National Ambient Air Quality Standards, 18
national parks, 32, 56
National Priorities List (NPL), 143, 148–49
National Toxics Inventory (NTI), 58
Natural Gas Star (EPA program), 103
Natural Resources Conservation Service, 133
Nelson, Gaylord, 13
Niagara Falls, 96
Niger Delta, 85–92
Nigeria, 83–84
Nikiforuk, Andrew, 26
nitrogen dioxide, 20, 54, 55
nitrogen oxide, 32, 48, 66
Nixon, Richard, 13, 139
nongovernmental organizations (NGOs), 91–92
Northeast Ozone Transport Region, 21

occupational hazards, 22
Occupational Safety and Health Administration (OSHA), 22
oil (fuel)
  air pollution and, 45
  economic and security risks, 46
  economy standards for, 46–47
  emission standards for, 46
  extraction of, in Niger Delta, 87–92
  hydrocarbons and, 49
  nitrogen oxides and, 48
  sulfur dioxide and, 48–49

see also oil spills and leaks
Oil Pollution Act (1990), 103
oil spills and leaks, 45–46, 80, 88
decrease in, 103–104
plastics production and, 177
outdoor air pollution, 18–22
ozone
depletion/pollution, 19
continued threat of, 38
decrease in, 55
EPA standards for, 137
global warming and, 42–43
health effects, 138
hydrocarbons and, 49
pesticides and, 40–42
phasing out chemicals for, 36–38
regional trends in levels of, 55–56
Ozone Transport Assessment Group (OTAG), 21
Ozone Transport Commission, 21

packaging, beverage, 94
paper industry, 175
paper recycling, 171–72, 178–80
particulate matter, 19, 47–48, 56–57, 138
Passacantando, John, 38
pentachlorophenol, 124, 125–26
perchloroethylene, 58, 59
Peru, 84
pesticides (methyl bromide), 40–42
petroleum transportation, 106
plastics, 105, 176–77, 204–205
Platt, Brenda, 182
Political Ecology Group (PEG), 40
pollution
environmental movement and, 12–13
landfills increase, 168–69
mines and, 177–78
oil extraction in Niger Delta and, 87–91
paper industry and, 175–76
paper recycling mills and, 178–80
plastics production and, 176–77
public awareness and concern for, 13–14
recycling transportation and, 180–81
reduction in, 13–14
by corporations, 94–96
by Mexican border companies, 98–100
public support for, 113–14
recycling helps, 174, 175
Pombo, Richard, 40
Porter, J. Winston, 189
Portney, Paul, 109, 112, 113, 159–60
price fixing, 79
Prince William Sound, Alaska, 80
Proctor and Gamble, 82
public opinion
concern for pollution, 13–14
on corporations, 77
on protecting the environment, 113–14

radon, 24
rail systems, 80
Reason Foundation, 166, 174, 178
recycling
benefits of, 182–83
car parts, 44
costs, 169–70, 191–92
cutbacks of curbside programs, 192
debate on, 164–65, 166–67
does not reduce waste, 203–204
as an economic development tool, 185–86
growth of, 164, 166, 182, 188–89
incentives/reasons for, 164, 202
is not convenient, 172–73
justifications for, misconceptions on, 189–90
market pricing and, 200–201
metals, 177
by Mexican border factories, 99, 100
mills for paper, 178–80
as neither good nor bad, 195–97
by oil and natural gas industry, 104–105
plastics, 177, 204–205
policing residents for, 193
problems with government policies on, 198–200
products from
companies for, 186–87
vs. manufacturing from virgin materials, 167–68, 197–98
market for, 170–71
vs. purchasing cheapest products, 172
questioning environmental benefits of, 190–91
reducing costs through, 184–85
reevaluating support for, 193–94
transportation for, 180–81
trees and, 171–72
wasting is outpacing rise in, 183
regulations
air quality standards, 137, 138–40
coal use, 64–65
corporations on, 109–10
enforcing/monitoring, 124
cycle of, 125
economics of, 128–30
vs. voluntary compliance, 125, 128
for wood preservation industry, 125–28
extreme costs of, 162
false assumptions for, 161–62

on indoor air pollution, 22
lack of demonstrable risks for, 157–58
mercury, 65
methyl bromide use, 40–41
need for continued, 109
for nonpoint water pollution, 16
partisan politics and, 110
public support for, 110, 113–14
scientific data for, 154–56
reducing emissions, debate on, 111–13
see also legislation; standards
Resource Conservation and Recovery Act (RCRA) (1976), 13, 116, 117
Reuter, Frederick, 160, 161
Rio Treaty (1997), 73
Ruckelshaus, William, 152–53
Ruston, John F., 166

Safe Drinking Water Act (1974), 134
Santer, Benjamin, 74
Scarlett, Lynn, 93, 195
Schindler, David, 29, 30
Schmolke, Richard, 39
scrubber technologies, 66–67
secondhand tobacco smoke, 23, 24
Seldman, Neil, 182
Shapiro, Robert, 81
Shaw, Jane S., 93
Shell, 83–84, 87, 89–90
Shenandoah National Park, 32
Sierra Club, 31
Silent Spring (Carson), 12
SIP Call rule, 64
Smith, Fred L., 70–71
Smithfield Packing Co., 16
smog, 32
soot, 32–33
South Dakota, 62
Southern Appalachian Mountains Initiative, 21
Spar, Brent, 83
Stafford, Robert, 118
Standard Oil of California, 80
standards
air pollution/quality, 17–18, 24–25, 53, 54
coal use, 31
emission, 46, 72–73
factory farm, 30
fuel economy, 46–47
State Revolving Loan Fund, 134
steel, 44, 94
steel mills, 67
stormwater runoff, 131
Stroup, Richard L., 157
sulfur dioxide, 20, 48–49, 56
Superfund program
cleanup process under, 143–44
as comprehensive, 118–19
creating new environmental habitats, 122–23
economics of, 97, 142
EPA goals in, 120–21
faulty risk assessments for, 146–47
as flawed, 141
future challenges to, 123
has not improved, 147–48
liability system for, 142, 144–46
need for reform of, 148–50
origin/creation of, 96, 117, 141–42
partnerships formed under, 121–22
trust fund for, 119–20

Tabazadeh, Azadeh, 41, 42–43
Taylor, Jerry, 165
tetrachlorophenol (TTCP), 126
tetraethyl lead, 18–19
Texaco, 87
Thomas, Bill, 42
Tierney, John, 167, 174, 178, 195
timber industry, 175
tobacco smoke, 23, 24
toxic substances/pollution
auto use and, 49–50
coal use and, 33–34
corporations and, 77
EPA monitoring air, 58–59
explosions, 116–17
incinerators and, 203
legislation on, 116
Love Canal, 96–97, 115
methyl bromide, 40–42
plastics, 176, 204–205
sources of air, 58
spills, 116
see also Superfund program
Toxic Substances Control Act (TSCA) (1976), 116
Train, Russell, 153
Transnational Resource and Action Center (TRAC), 40
trees, 171–72
TriCal, 40–41
TRW (Thompson, Ramo, Woolridge), 99

underground storage tanks (gasoline), 103–104
Union Carbide Corporation, 77, 79
Union Oil Company, 12
United States
Department of Agriculture (USDA)

Soil Conservation Service, 133
Water Quality Initiative (1989), 135
Department of Energy, 22, 67–68
Department of Health and Human Services, 25
Department of Justice, 79
General Accounting Office, 154–55
University of Guelph (Ontario, Canada), 62
Urban Ore, 186–87
utility companies, 65

Van Ray, Cor, 26
Vision 21, 68
volatile organic compounds, 23, 32, 49
Volokh, Alexander, 195

Walkerton, Ontario, 29
Waste Policy Center, 166
WasteWise partners, 184
water pollution
acid rain and, 33
factory farming and, 27–29, 60–62
lack of scientific analysis for assessing, 154–56
money spent on, 156
natural gas industry is helping, 103–104
nonpoint sources, 16, 131–34
recycling and, 179–80
scientific data on, 154–56
wood preservation industry and, 124–28
see also Clean Water Act; pollution
weather. *See* global warming
Weinberg, Adam S., 164
Werbe, Peter, 202
wood preservation, 124–28
World Meteorological Organization (WMO), 38
World Wildlife Fund (WWF), 82

zero waste, 183–84
Zuesse, Eric, 96